Half A Century
Of Photographing Railways

Michael Rhodes

PLATFORM 5

Published by Platform 5 Publishing Ltd, 52 Broadfield Road, Sheffield, S8 0XJ. England.

Printed in England by The Lavenham Press Ltd, Lavenham, Suffolk.
ISBN: 978 1909431 92 8

Contents

Front cover: One of the most iconic locations of the 21st Century was Wrawby Junction to the west of Barnetby. The busy freight traffic, coupled with the wonderful semaphore signals there, made it a mecca for photographers. On 3 February 2012, 66952 heads away from the photographer with an Immingham to Drax coal train. In the opposite direction a Manchester Airport to Cleethorpes express formed by 185132 arrives at Wrawby Junction on the line from Scunthorpe, whilst 66718 waits on the Brigg line with an empty coal train returning from Cottam to Immingham Docks. The junction was resignalled in December 2015 and shortly thereafter a massive levy was imposed on imported power station coal of around £38/ton, which led to the collapse of coal traffic through the port of Immingham. **Photographic Details: Canon EOS 7D, ISO 400, Canon 70-200mm f4L lens, 1/320 at f14.**

Back cover (left): Taken in 1974 this pictures a grumpy me at the end of Platform 6 & 7 at my beloved Cardiff General (now Central). The thin leather strap around my shoulder dates it the days of using my father's Agfa Silette. I have no idea who took it but for some reason I wanted a record of the day out!

Back Cover (right): Ever since the 1970s, I have had a fascination with signal boxes, so it was marvellous to spend much of 2016 working part time as a consultant surgeon at Stepping Hill Hospital in Stockport, because this was the closest hospital to the signal box at Great Rocks where my friend Alex Fisher is the resident signaller. On 17 August 2016, it looks as if I called in for an early breakfast before work with one of Alex's legendary breakfast rolls. **Alex Fisher. Photographic Details: Lumix TZ100, ISO 200, 9.1mm (25mm in 35mm), 1/20 at f5.**

Introduction

It has been both enjoyable and challenging to put this album together for Platform 5 Publishing. They say a picture speaks a thousand words and looking back through images that are up to 50 years old has brought many happy memories flooding back, hence the anecdotes and stories in many of the captions. The challenge has been that I find I have taken well over a quarter of a million railway related images since I first sallied forth in 1972. It was easy to keep track of things in the days I shot on film, as each roll of black and white was filed in folders and I knew I had taken a little over 50,000 images from 1972 to 2002. Similarly, all my slides were stored in folders and there were just under 30,000 of those.

The problem arose with the arrival of digital photography, and the "zero cost" of trying more views and taking more images. I haven't counted exactly, but I was running at an average of 3000 UK images each year from 2010, plus an average of 3000 images per trip for my many visits to China. Similarly, my work took me to the USA an average of four times a year and to Europe a little more often. I always took at least one camera with me when lecturing or presenting work overseas. Added to that were excellent holiday trips to Zimbabwe, Eritrea, Burma, Sri Lanka, South Africa and many other places with Farrail Tours.

The end result of trawling through all my images that are scanned and saved on several external hard drives is this book. There is a pattern to the photographs which is very much dictated by the "circle of life". Prior to 1978 I was a schoolboy in Cardiff with limited resources, but plenty of holidays and relatively few commitments, hence the many images taken in South Wales. Between 1978 and 1984, I was a student in Cambridge and then in Oxford studying medicine. Again, holidays were plentiful and with summer work in Germany, which actually started in 1977, resources were much better. Life as a junior doctor and while training as a general surgeon between 1984 and 1995, was spent mostly in the North East, hence the many images from that area.

In 1995 I was appointed a consultant surgeon in Norwich and there follows what I call my "Morrison Gap". I named this after the legendary photographer Brian Morrison, who once explained to us at a Phoenix Railway Photographic Circle gathering that 10–15 years of railway photography was missing from his collection because of the commitments of his work and family. In the same way, I found myself working as an NHS consultant, as well as doing private work and most importantly, supervising surgical research. When combined, these three aspects of life demanded at least 60 hours every week and I had three children too. The content of this collection reflects these commitments, with relatively few UK images taken between 1995 and 2010.

It's been said however that railway enthusiasm and railway photography are incurable diseases! During this 15-year period, I was privileged to lecture all over the world, as well as helping many research fellows present our research at academic meetings in Europe, the USA and further afield. I always took a camera with me and sought out the nearest railways. Perhaps the largest number of photographs were however, taken in China, where I visited for the second time in 1996 with two fellow photographers, the late Vernon Murphy and Steve LeCheminant. Travelling to China was an escape from the frenetic existence of a consultant surgeon and allowed me to switch my mobile phone off for a fortnight! By 2017, I had made a total of 33 trips to China specifically to photograph steam, visiting most of the provinces at some point and recording another 60,000 or so images of both steam railways and life in China.

As life took a quieter pace from 2010, and part time work became a possibility, my photographic output from UK travel increased dramatically. Not only that, but the opportunity to record all of the remaining mechanical and heritage signal boxes in the UK led to many days and weeks by the lineside. The results of this project is my 2020 book The Encyclopaedia of 21st Century Signal Boxes (Platform 5 Publishing). Finally, in 2018 I retired from surgery to concentrate on setting up a vineyard and expanding our small winery, transforming from a surgeon into a farmer. In the process, I have discovered, just as the excellent Clarkson's Farm series has so elegantly demonstrated, that this is far harder than it looks!

I know many world-renowned photographers pursued their hobby as a solitary endeavour, but for me it has always been predominantly about friendship and camaraderie. I would never have taken so many pictures as a student without the friendship and plotting of my fellow student Paul Shannon. Once working in the North East, David Allen in Durham always had ideas as to where one might have a day out. Similarly, Ken Groundwater suggested various trips. The perfect mix of the Extreme Steam Team (Rhodes, Murphy & Le Cheminant) where I was the Richard Hammond of the team, made every trip to China immensely enjoyable. Then later in life, the friendship of signallers and folk who bought some of our early books has led to visits and collaboration on new projects. Most recently was a new series called Freight Only – Germany, which was undertaken with an expert, Guy Houston, who bought my own Freight Only books as a teenager. It has also been great to travel with Bernd Seiler and his Farrail Tours, and meet regularly with similarly afflicted individuals – I am thinking of John Birkinshaw, Ralph Montague, John Parker, Trevor Maxted and many more who help me keep the bar car in business, whilst also trying to take a few decent pictures! Opportunities for international travel are reducing and the UK railway scene is quite frankly not as interesting as it used to be. But still the niggling feeling bubbles up after a week or two without taking any railway photographs, that I need "to get out with my camera"! I suspect it will always be so.

Michael Rhodes
Thurston, Suffolk
Summer 2022

Canon Power Shot Pro1

Above: From left to right are myself, Steve Le Cheminant, Alan Fengwang (our guide for over half of our China trips) and the late Vernon Murphy. We are pictured by our bus driver, using my Canon Power Shot Pro1, having climbed to the top of Hadashan mountain in September 2004. Perhaps the most photographically rewarding part of the last 50 years has been the many visits to China with two such excellent travelling companions. **Photographic Details: Canon Power Shot Pro1 at 11.5mm (64mm on 35mm), ISO100, 1/320 at f4.5.**

Below: 48 years into this epic voyage in photography and I am caught on camera by my fellow traveller Steve Arthur who is using his iphone 7 plus to show me "chimping" my shots at Dhurri Junction and showing them to the kind gentlemen who invited us onto the roof of his house to get a better view of the magnificent semaphore signals there. The train hasn't even finished passing us – what a contrast to the days and weeks waiting for films to be processed. **Steve Arthur. Photographic Details: iphone 7 plus, 4mm (28mm on 35mm), ISO 20, 1/550 at f1.8.**

A Photographic Journey

Looking back on my bookshelves, I find my first "Combined Volume" was the 1972 paperback edition with an AC electric locomotive on the cover. Inside the front page is the signature of the late Norman Preedy! I should perhaps explain how this came about. Whilst I had always been interested in railways since an early age and had a Triang model railway since the age of five, I despised diesels. It was not until a half term holiday in 1972, that I was persuaded by a school friend to take a visit to Cardiff General (as it was known at the time before it was renamed Cardiff Central in 1973) and to the sheds at Canton. The depot was packed with Westerns, Hymeks and even a couple of Warships. As we stood on the now-demolished footbridge from DeCroche Place, D1677 "THOR" thundered past with a Swansea to Paddington express and I was hooked. From then on, most Saturdays were spent on the platform at Cardiff General. On one afternoon, a small crowd had gathered round a "grown up" with a large camera (which I now know to be a Pentax 6x7), who was selling postcard photographs of locomotives. I purchased three, one each of a Western, a Hymek and a Warship. I was so impressed with the quality of the pictures compared to my disastrous efforts that I asked for the man's autograph and so my first combined volume came to be signed by Norman Preedy.

My photography started with a Kodak 126 instamatic. My pictures were always blurred and the negatives from the colour print film have long ago deteriorated so badly that only one image from those first months has been included in this book. The poor quality of my results, when compared to the postcards I had purchased, led me to nag my father incessantly to be allowed to use his Agfa Silette, a treasured family possession which had travelled with us all over North America in 1967 and 1968. He relented and my picture quality immediately improved. I took almost exclusively black and white film, which was processed at Boots and returned in an envelope with strips of negatives and postcard sized prints. Looking back, I suspect this was initially because black and white was significantly cheaper than colour film, but also because my father arranged for me to spend Saturday mornings in the medical illustration department at the newly opened University Hospital of Wales. It was here in Roger Marshall's department that I was taught how to process my own films and print in black and white. I was amazed at how much better my images were when developed and printed under the guidance of the professional photographers in the hospital.

Kodac Instamatic

The Agfa Silette didn't seem to be built for the intensive use it received once I commandeered it; the focus ring fell off and it began to fall apart. I then persuaded my father to lend me his Olympus Trip, but this was only allowed for special occasions, so most of my 1975 images were taken on days out or weeks away, rather than on my routine visits to Cardiff Central (as it had become). There still wasn't the ability to easily capture moving trains and I was frustrated by the fixed focal length of the lens, which was slightly too wide angle. I hatched a plan. I had seen a Praktica LTL with a 50mm f2.8 Tessar lens on offer for £49.95, which is equivalent to £400 using 2021's prices, in the local camera shop in May 1976. I persuaded my father to buy this on the basis that I could only have it if I achieved good grades at O-level. He agreed to buy it but sternly told me I couldn't have it until August when the results came out. Fortunately, I managed to persuade him after the exams finished in June 1976 that I had done well, so he relented and I was able to get the last summer of the Westerns on the sea wall with a decent camera. This was the summer of 76, the warmest and driest for a century and so the new camera got a lot of use before the results day came.

Agfa Silette

Olympus Trip

The next camera upgrade came from relentless reading of "Amateur Photographer", in which I kept looking at the new Canon AE-1. This was selling at £179.95, which equates to £1200 using 2021's prices and was therefore way beyond my budget, with just a paper round paying £1 per week. It was also way beyond my father's budget. However, another Rhodes family incentive scheme proved perfect to help me buy my treasured Canon in 1977. Uncertain as to whether to study engineering or medicine, my father intervened after I had chosen to study Maths, Physics and Chemistry, to mischievously suggest that if I also did A level German, he might be able to get me a summer job in Germany in July 1977. Realising that summer jobs in Germany paid handsomely in those days, I took the extra A level and earned enough that summer to buy an AE-1. Over the next few summers I worked in Germany during my school (and then university) holidays and was able to purchase new lenses. I had Praktica and Canon cameras, although many lenses were made by Tamron which offered an interchangeable lens mount. I found the coupling on the interchangeable lens mount to be unreliable and this was overcome in 1979, when I bought a Canon AT-1 and changed over to Canon lenses. My favourite lens by some margin was the lovely little Canon FD 100mm f2.8. One improvement with the AT-1 was the open aperture metering and the manual exposure which prevented much of the under-exposure that seemed to happen with the Canon AE-1. Little did I realise initially, but the automatic exposure mode on the AE-1 was very heavily

Practika LTL

Canon AE-1

Canon AT-1

centre weighted and the inclusion of just a fraction more sky, could reduce the exposure by 2 or 3 stops, resulting in quite a few under-exposed images for the first year or two.

With the initial struggles to afford decent camera equipment surmounted, I photographed all over the UK and quite a lot in Germany, until in 1984 I qualified as a doctor. The irony of life is that once one has a decent job and income, one rarely has much time to take photographs. This dilemma is beautifully described in The Big Picture by Douglas Kennedy. Here the protagonist is a New York lawyer who buys a Canon EOS 1N, only to find he never has time to take photos and his wife has an affair with a photographer in their suburban town. I shall say no more for fear of spoiling the plot. One notable extra purchase in 1985 was the Canon FD 300mm f5.6 which was light, pin sharp and probably my favourite lens of all time. It may sound macabre, but when working as a house officer in Worcester, I was paid by the mortuary for filling in death certificates. The proceeds of this work were paid to me in a brown envelope, which was pinned to the notice board in the doctor's mess, with the recipient's name on it. Those envelopes paid for this new lens which became my most used lens for the next five years.

Canon EOS-1N

Canon FD 300mm f5.6 lens

Whilst training as a surgeon over the next 11 years, we lived in Oxford, Worcester, Newcastle, Bristol, Brisbane and Cardiff and my photography very much reflected that. In Newcastle I became friendly with Ken in the camera shop on Gosforth High Street and it was he who persuaded me in 1991 to change my equipment to the new EOS autofocus system with the EOS-100 and then the much better EOS-10. I then stayed with this equipment until I went digital in 2003. Ken also persuaded me to dabble with medium format photography, where the negative size was much larger at 2¼ inch square. I had an old Yashica 124G which produced some superb quality negatives and slides, but was a nightmare to use on moving subjects. This was because the image was reversed in the viewing screen on top and the "sport finder", which was a metal square hole and the crudest of guides for framing a picture. Ken persuaded me to invest in a Mamiya 1000S which he had received to sell second hand. This was an excellent camera and much better for moving targets, although it was a bit on the heavy side.

Canon EOS 10

Yaschica 124G

Canon EOS 100

Mamiya 1000S

Once appointed a consultant surgeon in Norwich, I treated myself to a brace of EOS 1N camera bodies and found them to be exceptional, if a bit heavy with all the lenses that went with them. On trips to China from 1996 onwards, my cameras weighed over 20 kg and had to be carried in a large rucksack. That was fine in my 30s and 40s, but in my late 50s the weight of the equipment became a nuisance. Before I come to lightening my load, there was of course the change to digital. In its early years, the controversy about whether digital photography could ever match film was fierce and it was not until January 2003, when I received a phone call from Ken in Gosforth, that I took the plunge. Canon had allocated his shop just one EOS 1Ds 11.3 megapixel full frame camera body, which would be compatible with all the lenses in my collection. It cost an eye watering £6000 (equivalent to £10,500 using 2021's prices), but he assured me it was worth it. I was astonished by the quality of the prints from it after taking my Mamiya loaded with Fuji Provia film and the new EOS to Brundall station, where I took identical images of a Class 150 unit. The professional laboratory in Norwich produced two prints for me and comparison showed the digital print to be sharper, have better colours and a greater range of tones. I decided to start selling all my old film cameras, but kept one to load with black and white film, although looking back, why I did this I am not sure. A couple of years later the EOS 1Ds Mk ii came out and again after Ken phoned me, I bought it. The leap in quality was significant. A third camera was then added to the stable in unusual circumstances. Irene and I got married in December 2009 and a friend from New York attended the wedding. He called a few months before the ceremony to ostensibly ask for advice on which camera to buy for his wildlife photography. I suggested either the Canon 5D Mk ii or the new out 7D. He actually bought both and gave us the 7D as a wedding present. From that day, my wife has had a dislike of the 7D, complaining it wasn't much of a present for her!

Canon EOS 1D

Canon EOS 1Ds Mkii

Canon EOS 7D

Canon EOS 5D

Once appointed as a consultant surgeon in 1995, opportunities for photography in the UK were greatly diminished because of the commitments of a 60-hour week and a young family. The problem with railway enthusiasm and the love of photography though, is that both are lifelong passions. So it was from 1996 onwards that I embarked on regular trips to China to photograph the last great steam show on earth. Whenever I was in the UK, I was basically committed to work both in the NHS and in my private practice, so not more than the odd day out each year was possible. The advantage of taking a couple of weeks' annual leave to go to China was that one was free from receiving phone calls and other responsibilities. Consequently, the majority of my photographic output from 1996 to 2010 was in China. A second set of opportunities came from work related travel to present research and to teach or lecture. This was mainly in the USA, but also in Europe and it was usually possible to take a Saturday out and spend it by the lineside before flying home on a Sunday.

Since 2010, I was fortunate enough to work part-time, until full retirement from surgery in 2018. This allowed significantly more railway photography and the opportunity to travel purely for photography. Fantastic trips to Zimbabwe, Eritrea, Burma, Sri Lanka and South Africa in search of steam have been possible. Perhaps the biggest challenge I have undertaken however, has been to try and photograph the UK's surviving heritage signal boxes, with the help of many friends within Network Rail. Some of the images from this project have been included in this album.

Canon G15

Panasonic Lumix TZ100

By 2015, my camera equipment was becoming too heavy. I had tried purchasing various "pocket cameras"; for several years, the Canon G15 was my favourite, but the limitations of the small sensor were obvious above 400 ASA. Another excellent pocket camera, the Panasonic Lumix TZ100 was great when I was commuting by plane on a weekly basis from Norwich to either Manchester or Exeter for work; it fitted in my hand baggage on the tiny propellor planes and yielded excellent images with the caveat that the zoom lens became a little soft above 200mm. I still have the remarkable Sony RX100 Mk vii for trips where a camera bag is too large. It produces amazing results and is fast and responsive with an excellent 24–200mm equivalent zoom lens.

Olympus EM1

One day in 2015, with camera weight on my mind, I wandered into the Campkins camera shop on Kings Parade in Cambridge and the assistant showed me a series of prints taken with the new Olympus EM1. The quality was superb and the lenses with the camera turned out to be as sharp as anything produced by Canon. I refer specifically to the amazing 40–150mm f2.8 and the 75mm f1.8. For those interested, there are many detailed reviews on the internet, but all I can say is that in real life, the lenses were a revelation. And so, after nearly 40 years since 1977, I sold all my Canon lenses and cameras and switched to Olympus. The system weighed half as much and the selection of lenses was superb. I would still be using the Olympus, but for the fact that on a trip to Germany in October 2019, our hire car was broken into and all the camera equipment in the boot stolen. Faced with trying to replace almost £10,000-worth of equipment, including my new DJI drone and finding that none of it was covered by the car insurance, travel insurance or home insurance, I sat back and wondered about just using my surviving little Sony RX100 Mk vii.

Sony RX100 Mk vii

Challenges of a different kind began in March 2020, when we were all locked down and but for a few months respite in summer 2020, we have had very little opportunity to travel overseas and only limited opportunities to travel for leisure in the UK. During one outing with my good friends Steve Arthur and Martyn Fordham, Martyn spent much of the day extoling the virtues of his new Sony A7R Mk iv, especially its amazing dynamic range. I researched the camera, looked at www.dpreview.com, which is my go to website for clear and honest reviews, and took the plunge and bought one. It may sound trivial, but Sony zooms turn the opposite way to Canon and Olympus zooms and after over 40 years zooming one way, I found myself too old and grumpy to adapt and sold it within a few months. I've now gone back to Olympus and am saving to gradually replace my stolen gear.

Two final aspects of this photographic journey are the introduction of drones and the amazing improvement in smartphone cameras. First drones; the DJI drones are quite a revelation. Having spent much of my youth scaling lighting masts to get better views of trains and yards, and often dreaming of going up in a helicopter, drones have now arrived. The images selected for the last few years of this collection, reflect the new perspectives and opportunities that these amazing machines open up. Then the smart phone. The new iphone 12 pro max has three lenses and on a sunny day a standard lens photograph can be printed at A3 size and is indistinguishable from a digital image taken in 2003 on the £6000 EOS 1Ds. The improvement in the iphone is almost a piece of science fiction and when it comes to family and friends, all my photographs are now taken on an iphone. I have even had a couple of railway images taken with an iphone published and looking at the magazine, it is impossible to tell that these were taken using a phone.

So what of the future? At the time of writing in 2021, things are a bit depressing. International travel looks like being a struggle for some time to come. A recent trip to Germany required PDFs of traveller's vaccination status, locator forms, a lateral flow test two days prior to returning to the UK, another passenger locator form for the UK with an uploaded negative lateral flow test and then a PCR test two days after one's return. No wonder Stansted airport was deserted for both the outbound and return journey. It feels as though one needs a PhD in logistics just to cross the Channel and then one has to queue in the non-EU line and gets a passport stamp, just for a trip to Germany. In addition, potential trips can be sabotaged at a few days' notice by changes in the regulations, as has happened to me several times over the last 18 months. At home, it is difficult having visited almost every signal box and most locations in the UK, to generate enthusiasm for a long drive and a day out when one has already taken pictures at the locations to be visited. Time will tell how things change over the next decade, but it has certainly been a fantastic photographic adventure over the last 50 years and I wouldn't have missed it for the world.

Sony A7R Mk iv and 24–105mm zoom lens

The Olympus OM-D E-M1 Mk iii which I currently use.

Photographic Details

As photography became the most important aspect of my interest in railways, I started to record details of the camera equipment, film and settings that were used for each image that I took. Those records have allowed me to provide the reader with information about the equipment and settings for each of the images illustrated. In the digital era, which for me started in 2003, the EXIF data that is embedded into each image provides a 100% accurate record of how each image was recorded. For the digital images, instead of the film used, details of the ISO setting are given. This information is beneath each image and is laid out as follows:

Camera used, film used, lens details, zoom (if applicable), shutter speed in seconds, aperture.
E.g. Canon AE-1, Ektachrome 200, Tamron 200mm f3.5 lens, 1/250 at f5.6.

The 1970s

1972 to 1974

These three years were spent at school in Cardiff. The limitations of trying to take pictures as a schoolboy earning £1 per week on his paper round are evident in these images. But what I lacked in photographic prowess, I more than made up for in number collecting, travelling the length and breadth of the country visiting locomotive depots and hanging around on station platforms, sitting in BRUTE (British Rail Universal Trolley Equipment) parcels trolleys, sipping lemonade and eating home-made sandwiches.

Above: This is one of my earliest images and was taken some time during 1972. As I recall there was an 09.00 departure from Cardiff for Paddington and an 09.05 to Portsmouth Harbour. The train reporting numbers displayed on the locomotives' headcode panels confirm that the "Hymek" on the left is destined for Portsmouth Harbour and the "Western" on the right is destined for London Paddington. This combination of motive power could often be seen on Platforms 1 and 2 at Cardiff General (as it was called in those days). **Photographic Details: Kodak Instamatic 126, Boots colour film and printing.**

Right: An atrocious, grainy, scratched image from the first time I borrowed my father's Agfa Silette. It does however, remind me of a caper that would be regarded as child neglect today. It was Easter 1973 and I was not yet 13. I had persuaded my parents that me and my school friend "Piff" would be safe to travel from Cardiff to my grandparents in Great Harwood near Blackburn and that we could go trainspotting each day, either on the bus to Manchester, or on the train to Preston. On the Saturday of our stay, we went to Preston bright and early, with extra cash hidden in our pockets. Why? Just weeks earlier, the new electric service from Preston to Glasgow had been introduced and we'd hatched a plan to pretend to be in Preston all day, but head to Glasgow and Edinburgh! I was not quite five feet tall at the time but my friend Piff was much taller and looked at least 15. The caper found us on the Cowlairs Road at 14.00 on a Saturday, looking for Eastfield depot, with me asking why the policeman was leaving the men on the other side of the road to punch each other without arresting them – his answer was unintelligible but along the lines, let them tire themselves out! This image was from the first leg of our journey. We had changed at Carstairs to head to Edinburgh and D5065 (which would become 24065 under TOPS) was on the shorter-distance connecting train from Carstairs to Edinburgh. The Type 2 had failed though and English Electric Type 1 D8113 had been sent from Carstairs shed to take us to Edinburgh. We arrived back at my grandparents after 9pm, exhausted, but not questioned and were simply treated to my grandad's excellent egg, ham and chips, all washed down with a mug of tea! **Photographic Details: Agfa Silette, Boots Panchromatic X, home developed, 1/60 at f4.**

Below right: An unusually sharp picture (for me at the time) of English Electric Type 4 D347 passing through Sheffield Midland station in 1973. We often purchased a day return from Cardiff to Sheffield on a Saturday, as we worked out that we could visit Tinsley, Toton and one other depot before the last train home from Birmingham New Street to Cardiff. The third depot could be Burton upon Trent, Bescot or Saltley. These days out yielded huge numbers of "cops", but didn't produce many good photographs, as we were invariably "bunking" the depots, moving quickly in order to collect numbers rather than to take photographs. If only I had taken more pictures. **Photographic Details: Agfa Silette, Boots Panchromatic X, home developed, 1/125 at f5.6.**

Below right: I am not sure how many times I sat on the platform at Cardiff between 1972 and getting my first moped in the spring of 1977, but it must be in the hundreds. There was fun, excitement and even the occasional life lesson. In 1974, I was with a friend one winter's day and we each had a shilling (I know it was decimal by then and I should say 5p, but we still spoke in "old money"). It was freezing and we wanted to buy a cup of tea each and a sausage roll to share from the buffet, but we didn't quite have enough. My friend who regarded himself as worldly wise explained to me that he was an expert of the fruit machines in the buffet and he always won, so if I gave him my 5p he would double or triple our money and we could have a sausage roll each. Being from a religious family and having never gambled, I hesitated, but he assured me and so I gave him my coin. Of course, the expected happened and he lost it, then he lost his own money and we went hungry and thirsty! Anyway, one regular sight at the time was 4E80, the Danygraig to Stratford freightliner which was sometimes hauled by a Western. Here it is caught passing what I believe is the 17.37 Cardiff–Manchester, with a Class 2 passenger train, probably to Bristol, on the far right. Try as I might, I cannot find the locomotive numbers with any certainty, as my record keeping at the time was poor. **Photographic Details: Agfa Silette, Boots Panchromatic X, 1/125 at f3.5.**

Left: The better quality of my father's Olympus Trip is immediately evident in this view of 71013 stabled at Ashford in Kent. I recall that I managed to use the camera because my father was attending a British Society of Gastroenterology meeting which fell during a school holiday. I accompanied him to the meeting in Canterbury and immediately bunked off to travel to Dover and Ashford to try and see some Class 71s. **Photographic Details: Olympus Trip, Ilford FP4 at f4 with automatic shutter speed.**

Above: In July 1974 I stayed with a family friend who was a doctor in Sheffield and this allowed visits to what were a variety of new locomotive depots for me, including Wath, Doncaster, Frodingham, Immingham, York, Holbeck, Healey Mills, Shirebrook and Barrow Hill. Every day ended with a bus ride to Brinsworth and Three Magpies, and a ten-minute walk up the hill to Tinsley. On 24 July 1974, Class 20s 20204, 20023, 20215, 20003, 20133, 20201, 20203, 20058 & 20091 are seen at Tinsley in the evening sun. **Photographic Details: Agfa Silette, Ilford FP4, 1/125 at f8.**

Below right: Although Ebbw Junction was not far from our family home in Cardiff, I rarely visited because it was so difficult to get to from Newport station. On top of that, travelling back and forth between Cardiff and Newport with a day return allowed most locomotives stabled there to be recorded, as the depot was adjacent to the main line. So in 1974, when we worked out that we needed some of the Class 08s allocated to the depot, a friend and I plotted an evening trip, undertaken on Saturday 10 August 1974. It took us two hours to get from Newport station to the depot and back, but we were rewarded with seeing 37 locomotives on shed. Pictured here is the iconic experimental locomotive D1200 "Falcon" with the 9E61 headcode it carried for much of its working life, as it tripped between Llanwern steelworks, the Uskmouth branch and Newport Docks. Also seen are 25161 (one of the Class 25s allocated to Newport for local steel traffic at the time) and 47365, an interloper from "up north". The full list of locomotives present that day was: 08029, 08109, 08122, 08135, 08582, 08595, 08647, 08657, 08846, 08848, 08853, 08894, 08940, 25030, 25031, 25059, 25063, 25153, 25156, 25157, 25158, 25159, 25161, 25163, 25165, 25166, 25219, 37203, 37220, 37222, 37231, 37233, 37284, 37290, 37308, 47365 & D1200. **Photographic Details: Agfa Silette, Ilford FP4, 1/125 at f8.**

1975

1975 carried on much as the previous three years had, with the exception that O levels came on the horizon. My photographic darkroom skills improved (a bit) and I finally tried a couple of rolls of colour slide film.

Above: This image of 40050 passing through the magnificent station at York was taken during the summer of 1975. It represents results that are just about as good as I could manage with the Agfa Silette, its maximum shutter speed of 1/125 and my home processing. **Photographic Details: Agfa Silette, Boots Panchromatic X, 1/125 at f5.6.**

Right: We knew the Westerns were disappearing fast during 1975, so every opportunity was taken to catch a portrait of these locomotives on the Paddington to Swansea trains. Here, D1037 "WESTERN EMPRESS" stands on Platform 3 at Cardiff General in the summer of 1975, as a young spotter asks if he can cab the locomotive. My recollection is that cabbing a locomotive, as well as being hauled by it, gave extra kudos to the entry in the enthusiast's combined volume! **Photographic Details: Agfa Silette, Boots Panchromatic X, 1/125 at f8.**

Above: 1975 was the first year I took any colour transparencies and my first film was a roll of Agfa CT18. Everybody I asked told me that Kodachrome 25 ASA was the best film, but that it was very slow and the 64 ASA version was the next best thing. This however, was very expensive and came with a pre-paid processing envelope, which added to the cost. Consequently, Agfa and the atrocious Orwochrome, made in East Germany, were initially my slide film of preference, until I accepted that the results were so bad that I changed to Fujichrome R100, which was processed locally. For special trips, a roll of Kodachrome was sometimes purchased, but its greater cost meant that it had to last several months! This image of 55001 "ST. PADDY" at the stabling point just north of London Kings Cross ended up being significant, because although I took many hundreds of images of Deltics in the ensuing years, it proved to be my only picture of "ST. PADDY". **Photographic Details: Agfa Silette, Agfa CT18, 1/60 at f5.6.**

Left: I made my first visit to Germany in December 1975. Travel was by train from Cardiff to Paddington, then from Liverpool Street to Harwich, followed by the overnight ferry to Hoek van Holland and the "boat train" to Cologne. From there I took the train to Siegen where my pen friend Eberhards's family collected me, completing the 24-hour journey. I was allowed to take the Olympus Trip and a roll of slide film. This view of a derelict German 2-10-0 steam locomotive on Betzdorf depot typifies my attempts to capture working German steam. I had been told by a lad in the village of Holzhausen where I was staying, that the evening parcels train from Wurgendorf to Betzdorf was always steam-hauled and so we persuaded his dad to take us to the station one evening; it was a diesel and the driver told us that steam had finished "yesterday". I then persuaded my hosts to drive to Betzdorf where the steam locomotives were said to be allocated, only to find that steam had indeed finished that week. History repeated itself in 1977, when working in Germany, I made a visit to catch the Rheine to Emden steam spectacle, only to find it too had finished "the week before"! **Photographic Details: Olympus Trip, Agfa CT18 on automatic.**

1976

This was my O level year and also the year that I finally got an SLR in the form of a Practika LTL. This led to a significant improvement in the quality of my photographs, which I am sure is a relief to the reader.

Above: 1976 was the last summer that the Class 52 "Westerns" operated and in order to capture as many of them as possible, I made a summer Saturday trip to Exeter St. Davids on 10 July 1976. So you ask, why is this not an image of a Western? I have included this image because it reflects two other aspects of those frantic summer Saturdays in the West Country – freight traction on passenger services and milk trains. 47216, which was based at Immingham at the time, is at the head of the 08.45 Liverpool–Penzance, which it worked as far as Plymouth. On summer Saturdays, the Class 47s that were usually employed on freight workings in the north could often be found on passenger services. 25220 stands alongside the passenger train with a rake of empty milk tanks running from Acton Yard to Lostwithiel. This and the Carmarthen milk tanks were the last milk services on the rail network, with the exception of the ill-fated Milk Liner (see page 103). **Photographic Details: Practika LTL, Boots Panchromatic X, Tessar 50mm f2.8 lens, 1/250 at f8.**

Below: My most memorable trip of the summer of 1976, the hottest on record, was a one-week Scottish Region rail rover. On Saturday 14 August, there were a remarkable 70 locomotives and 25 DMU vehicles to be found on the depot at Eastfield, some of which are seen in this view. **Photographic Details: Practika LTL, Tessar 50mm f2.8 lens, 1/250 at f5.6.**

Above: In 1976, our church summer camp took place in Colwyn Bay. For anybody wishing to clear British Railways (that is, to see every locomotive on the network), a visit to the breakwater at Holyhead was mandatory. Why? Because there resided the two surviving Class 01s, which were located on rails that were not connected to the main line network. Just five of these small shunters were built and the other three had been withdrawn during the 1960s. This seemed too good an opportunity to miss, so one afternoon I bunked off and captured both locomotives on what was my first roll of Kodak TriX 400 ASA film. I never truly mastered the processing of this film as this image perhaps reveals. **Photographic Details: Practika LTL, Tessar 50mm f2.8 lens, 1/250 at f11.**

Below: And so to the last summer of the Westerns. I managed to buy my first roll of Kodachrome 64 ASA in July 1976 and spent 24 July 1976 on the sea wall at Dawlish with my Mum, Dad and three siblings. To be honest, not much time was spent with them, as I scuttled back and forth trying to predict which train would have a Western on the front. I think the film ran out part way through the day and I had to revert to black and white. I remember that I seemed to always be in the wrong spot for a Western, or the sun disappeared just as one approached. The sun went in as D1058 "WESTERN NOBLEMAN" roared past on an Up express, necessitating using the Tessar lens wide open. The lack of depth of field is noticeable, but at least I got one colour image of a Western on the sea wall.
Photographic Details: Practika LTL, Tessar 500mm f2.8 lens, 1/500 at f2.8.

1977

Reviewing my notebooks from 1977, it seems that my long suffering Godparents, who lived in Cheadle Hulme in South Manchester, had me to stay both in January and April 1977. In addition, I visited friends who were studying in Cambridge; they invited me to stay for a weekend to experience student life, which to me seemed all about drinking copious amounts of coffee and arguing about the meaning of life all night! The summer was entirely taken up by six weeks working in Germany at Boeringer Ingelheim as a market research assistant. Perhaps the biggest single impact on my photography, was the purchase of an elderly Honda C70 moped. This allowed me to explore the Welsh valleys and to chase some of the last special trains which commemorated the end of the working lives of the Westerns. A second major impact was the publication of the first edition of S.K. Baker's Rail Atlas of Britain. This enabled me to see for the first time where the freight only lines were. Needless to say, this had an impact on my photography not just in 1977, but for the next two decades, as I tried to visit every freight line. These highlights are reflected in the images I have selected for 1977, which was also the year that I took what would be my first photographs to be published.

Above: On the way back from Toton to Manchester, I decided to visit a depot I had never photographed before, Barrow Hill. I had been once back in 1974 but didn't take any pictures. The unusual angle of this shot is due to having to rest my camera on an oil barrel (I never had a tripod). From left to right, 20131, 47307, 08267, 20007 & 47281 are captured around the turntable inside the depot's roundhouse. **Photographic Details: Practika LTL, Tessar 50mm f2.8 lens, 1s at f5.6.**

Right: This image was published in the 1979 Locoshed Directory and remains a favourite of mine. On 6 January 1977, 76016 is being repaired in Reddish depot, along with 40096, 25115 and AC electric multiple unit 304005 which are also receiving attention. **Photographic Details: Practika LTL, Boots Panchromatic X, Tessar 50mm f2.8 lens, 1/30 at f5.6.**

Top right: Whilst staying in Manchester, I visited Deby Works for its open day on Saturday 8 January 1977, recording a total of 78 locomotives there. The works visit was followed by the obligatory Saturday trip to Toton, which as I recall was crawling with "spotters". There were 96 locomotives on shed, with examples of Classes 20, 31, 44, 45, 47 seen in this very gloomy afternoon view. **Photographic Details: Practika LTL, Boots Panchromatic X, Tessar 50mm f2.8 lens, 1/125 at f2.8.**

76016

Right: 26 February 1977 was a red letter day for railway enthusiasts in Cardiff. Not only did we get to see D1013 and D1023 pass through twice on The Western Tribute farewell railtour, but two different Class 40s arrived from Liverpool Lime Street; 40028 and 40120 both worked football specials in conjunction with the FA cup tie between Cardiff City and Everton. Cardiff City lost. Here, D1013 "WESTERN RANGER" and D1023 "WESTERN FUSILIER" are seen arriving, with the Swansea to Plymouth leg of their final journey back to London. This image was one of the first I shot using my newly acquired Pentacon 135mm lens, which had a manual diaphragm that had to be shut down by hand before taking a picture; it was very cheap, but produced sharp results! **Photographic Details: Practika LTL, Boots Panchromatic X, Pentacon 135mm f2.8 lens, 1/250 at f8.**

Below: A couple of weeks earlier, on 13 February 1977, the Western Requiem railtour ran on the Welsh valley lines to Treherbert, Merthyr Tydfil and Aberdare. D1010 "WESTERN CAMPAIGNER" was captured as it returns from Merthyr Tydfil and passes the signal box at Pontypridd. This was one of my first outings on my newly acquired Honda C70 and I chased the train south to Radyr for a second photograph. **Photographic Details: Practika LTL, Boots Panchromatic X, Tessar 50mm f2.8 lens, 1/250 at f4.**

Above: The newly purchased Pentacon 135 mm was also used to good effect during my visit to Cambridge, to discover what university could be like. On 19 April 1977, 37037 passes Cambridge North signal box with one of the two-hourly Kings Lynn to Liverpool Street expresses. **Photographic Details: Practika LTL, Boots Panchromatic X, Pentacon 135mm f2.8 lens, 1/250 at f5.6.**

Above: On 9 June 1977, 55007 "PINZA" is seen at Holgate Junction south of York with The Silver Jubilee, which was a named passenger train to mark the namesake milestone achieved by The Queen. Departing Edinburgh at 15.00 for London Kings Cross, this was only the second day that it ran. The image is especially poignant in retrospect as I photographed Her Majesty the Queen boarding her Diamond Jubilee train in Middleton Towers 35 years later. On the day in 1977, it was the mixed freight trains that seemed more interesting! **Photographic Details: Practika LTL, Boots Panchromatic X, Tessar 50mm f2.8 lens, 1/250 at f5.6.**

Below: There are many layers to the selection of this image, which has been included primarily because it is one of my favourite views, but also because it was taken from a location that was rarely visited by other photographers. On 13 July 1977, I was on my way to work in Germany and spent a couple of hours at St. Pancras while travelling from Cardiff to Harwich, where I would catch the ferry. I had a new lens to try for the trip, a Tamron 70–150mm f3.5–f3.8 interchangeable mount zoom. Purchased second hand, I had been assured it was the best on the market! I had decided that I had neglected St. Pancras during my many visits to London during the previous five years and so I spent a couple of hours trying to capture the Barlow-designed trainshed. As I recollect, I simply wandered off the platform and behind the signal box, having slipped on a yellow vest which I kept in my camera bag. Nobody batted an eyelid and I took half a dozen images from this spot before walking back onto the platform. **Photographic Details: Canon AE-1, Kodak TriX, Tamron 70–150mm f3.5 at 150mm lens, 1/250 at f5.6.**

Above: During a June 1977 trip at the end of school half term, I spent some time in Leeds and decided to try and get to the top floor of the newly constructed Dragonara Hotel. I wandered in and up to the top floor, where I found a vantage point to capture this view of 45127 arriving with a St. Pancras to Leeds service, just as 45118 leaves with a train to St. Pancras. The image has always been one I was pleased with, so in 2019 when attending an Association of Laparoscopic Surgeons meeting in Leeds, I decided to revisit the Dragonara Hotel (now the Hilton Leeds City) with a view to repeating the shot 42 years later. I took a print of the original, which the manager found fascinating and so I was taken to the same vantage point, only to discover that the newer Double Tree by Hilton blocked most of the view. All was not lost though, as the Double Tree has a roof top bar. I then discovered that the overall roof of Leeds station has been so expanded that the shot is almost unrecognisable. I took a picture of an Azuma, but in my mind the 1977 view is much better – I must be getting old! **Photographic Details: Practika LTL, Ilford FP4, Pentacon 135mm f2.8 lens, 1/250 at f8.**

Above: I think this was the first image that I had published. It was part of a March 1978 article by Geoffrey Freeman-Allen on how the German Railways were sinking deeper into debt. On 31 August 1977, I was on my way back from working in Ingelheim and had inserted the first roll of film into my Canon AE-1 that I had purchased a couple of hours earlier in the camera shop at Cologne Hauptbahnhof (Hbf). I had made time in my journey to allow all day on Hbf, enjoying the obligatory "Currywurst mit Pommes" and a can of beer. The Class 103 electrics were striking, in their red and cream livery, and every hour there was the chance of seeing three alongside each other as passengers interchanged between three southbound inter-city services. In 1977, freight lines in Germany were as much a mystery to me as they were in the UK until the first publication of S. K. Baker's rail atlas. That changed in 1994 when the Schweers & Wall atlas was published, revealing the extent of the freight only network in the vicinity of Cologne to photographers such as myself. **Photographic Details: Canon AE-1, Ilford FP4, Canon 50mm f1.8 lens, 1/250 at f5.6.**

Right: This "synchronised dismount" was not posed. The drivers of 08593 and 37285 climb down from their cabs in perfect harmony. The phurnacite plant at Abercwmboi was one of my earliest discoveries when researching the Welsh valleys using my Baker rail atlas, which was carried in my camera bag. The signal box there was permanently covered in grime from the coking plant, which provided significant amounts of rail traffic. A real bonus was that until 1980, the coal delivered there from Mountain Ash and Perhiwceiber collieries was steam hauled. The Class 08 was the Aberdare area pilot, whilst 37285 was putting together the afternoon trip working to Radyr yard. **Photographic Details: Canon AE-1, Kodak TriX, Tamron 70–150mm lens zoom at 150mm, 1/250 at f5.6.**

1978

1978 was an eventful year. Just 13 days before my A level exams started, during a break in revision in May 1978, I was knocked off my Honda moped on my way back from photographing 40182 (I had been tipped off that it was stabled on Canton shed). The resulting broken shoulder and knee meant a break in photography, but more seriously, having to sit my A levels with an amanuensis (a person you dictate to). I had broken my left shoulder and I am left-handed! My photographic output for the year was limited to one big summer trip to the North East and some days out, including a memorable moped ride over the Heads of the Valleys road to the amazing colliery railways at Brynlliw and Graig Merthyr. The latter part of the year was spent settling into university life in Cambridge and so intense was the experience that I appear to have hardly taken any pictures between September and the end of the year.

Above: A favourite spot for railway photographers is the castle keep north of Newcastle Central station. On 7 July 1978, 46032 is perhaps unusual motive power for an unfitted coal train from Low Fell Yard to Blyth Cambois Power Station. Coal trains from County Durham often stabled at Low Fell, which is two miles south of Newcastle, before completing the journey to Cambois Power Station, but these were usually hauled by a Class 37. **Photographic Details: Canon AE-1, Ilford FP4, Canon 50mm f1.8 lens, 1/250 at f8.**

Above: The church annual holiday was once again in Colwyn Bay and this time several of us engineered a day out to Chester; the girls to go shopping and two of us to record the semaphores and signal boxes in Chester! 40127 passes the magnificent Chester No. 2 signal box with an unidentified freight working which is arriving on the line from Crewe. Indeed, none other than my university friend Paul Shannon has said of this train "I've never seen a working like that at Chester". No. 2 signal box has since been demolished. It was the largest signal box at Chester and with just over 180 levers, it was the same size as the world's largest remaining lever box at Shrewsbury's Severn Bridge Junction. **Photographic details: Canon AE-1, Ilford FP4, Canon 100mm f2.8 lens, 1/250 at f4.**

Below: My only good image of hump shunting in the "down" yard at Severn Tunnel Junction was taken on 27 June 1978, which was the week after my A level exams finished and remarkably, only a few weeks after my moped accident. A friend gave me a lift to Severn Tunnel Junction and I had recovered from my fractures enough to capture 08940 in action. Over the decades, I have spent hundreds of hours on the long road bridge over the western end of Severn Tunnel Junction station. **Photographic Details: Canon AE-1, Ilford FP4, Tamron 70–150mm lens zoom at 100mm, 1/250 at f5.6.**

Above: The expedition to record the last days of steam at Graig Merthyr and Brinlliw collieries was undertaken on 6 April 1978. I arrived at the hillside overlooking the branch up to Graig Merthyr Colliery to find two photographers on a Honda 750cc motor bike which somewhat dwarfed my Honda 70 moped. One was the well-known photographer Kevin Lane, who I believe has an identical picture to this. Later in the colliery, the erratic automatic exposure of my Canon AE-1 led to the white steam rendering the rest of my slides three stops under-exposed, but at least I managed one decent image of this amazing branch line. **Photographic Details: Canon AE-1, Kodachrome 64, Canon 50mm f1.8 lens, 1/250 at f4.5.**

Left: Having recorded the pits to the west of Llanelli, I set off home via the Heads of the Valleys road, to avoid the motorway. Dropping down from Hirwaun, I had to pass the colliery system at Mountain Ash, but had run out of colour film. I met another well-known photographer there and Joe kindly sold me a roll of film, allowing me to capture this shot of the Hunslet Austerity 0-6-0ST (LNER Class J94) No. 8 crossing the bridge linking the two halves of the colliery system. This bridge was washed away in the floods of February 1980, leading to the premature closure of the colliery system at Mountain Ash. **Photographic Details: Canon AE-1, Kodachrome64, Canon 50mm f1.8 lens, 1/250 at f5.6.**

1979

After settling into university life, a few more visits were taken to the bridges at Hills Road and Mills Road in Cambridge, to capture the magnificent semaphores there. One other notable trip was an opportunist visit to Mountain Ash one freezing February morning on the way to a dental appointment at my cousin's surgery in Porth. Sadly, my camera was loaded with 200 ASA Ektachrome, which wasn't the best film, but the strongest two images from that wonderful morning are included in this year's selection. The summer was again spent working in Germany as a bookkeeper in a furniture company and much of the earnings were spent on new lenses and a three-week German rail rover.

Above: During my German rail rover, I visited the newly opened Hamburg Maschen yard, which was touted as the biggest in Europe and the busiest in the world. The long road bridge across the southern end of the yard was open in those days and I walked from the Maschen S-bahn station. As I was taking pictures from the bridge, a DB employee approached me and asked if I would like a tour of the yard. The yard was undoubtedly the jewel in the crown of the Deutsche Bahn at that time and its employees were very proud to work there. This image was taken in the Maschen SW signal box, which controls the exit from the N-S classification tracks. **Photographic Details: Canon AE-1, Ilford FP4, Tamron 28mm f2.8 lens, 1/30 at f5.6.**

Top right: Similar hospitality was shown by DB staff the day after my Maschen visit, when the driver of DB Class 220, 220 011 invited me into the cab to ride along with him from Cuxhaven to Bremerhaven. This view, taken on 2 August 1979 as we entered Wremen station, shows classmate 220 010 travelling in the other direction with a Bremerhaven to Cuxhaven local service. **Photographic Details: Canon AE-1, Ilford FP4, Tamron 28mm f2.8 lens, 1/125 at f4.**

Right: One final image from my German rail rover. Taken on 4 August 1979, this shows one of the turntables at Wurzburg depot. My notebook records that there were 52 locomotives on shed that day. Those around the large turntable included vintage Class 118 and Class 194 electrics, which were soon to disappear from the network. From left to right, the locos visible are 140 124, 194 594, 110 293, 111 098, 151 039, 118 013, 140 842, 140 765, 118 025, 118 045 and 141 237. **Photographic Details: Practika LTL, Fujichrome R100, Tamron 28mm f2.8 lens, 1/250 at f8.**

DB
220 010-3

Left: The significance of this image is that it won first prize in the Cambridge University Railway Club's 1980 photographic competition. This was judged by Rex Kennedy from the Oxford Publishing Company. In a conversation after the competition, Rex said that he'd be happy to consider publishing a book illustrating freight trains and thus "Freight Trains of British Rail" was commissioned; no contract, no paperwork, just a gentlemanly conversation and this was published in 1982. It was my first book and the rest as they say is history. The image shows 45077 looped at Wellingborough, with a Northfleet to Toton cement train, as 45128 speeds through the station on a Nottingham to St. Pancras express on 26 July 1979. **Photographic Details: Canon AE-1, Ilford FP4, Canon 100mm f2.8 lens, 1/250 at f5.6.**

Left: On 24 January 1979, prior to lectures starting I cycled down to Hills Road bridge in Cambridge. A Class 104, consisting of vehicles 50578, 59218 & 50585, departed with a stopping service to Bishops Stortford as I recall. The Metropolitan-Cammell DMU in the bay platform above this and which is partially obscured by the lamp posts, formed a Hitchin service, where passengers could change for Kings Cross. The 128-lever Cambridge South signal box was to close on 17 October 1982, along with Cambridge North box, which is visible in the background (72 levers) and the nearby Coldham Lane Junction signal box (101 levers). The re-signalling was a prelude to the forthcoming electrification of the route. **Photographic Details: Canon AE-1, Ilford FP4, Tamron 200mm f3.5 lens, 1/250 at f8.**

Above: The last three images in this 1979 selection are testimony to both the damage wrought on slides sent for publication before digital scanning was available and the recovery techniques that are now possible with good scanners and software. This image was taken at dawn on 14 December 1979, as the locomotives at Mountain Ash colliery were being prepared for the day's work. It was published in a magazine during the 1980s and in the process, the transparency was pasted to a drum scanner and was completely trashed. Indeed, so bad was the damage that the experience led me to cease submitting colour work for publication. Fast forward to 2020 and using a Nikon Super Coolscan 5000 and the vuescan software, all the blemishes were removed to recover what was one of my most evocative images on the Mountain Ash system. **Photographic Details: Canon AE-1, Ektachrome 200, Canon 100mm f2.8 lens, 1/250 at f11.**

Left: A second image, captured later on the 14 December 1979 after my dental check-up in Porth, sees No. 8 and Haulwen working a coal train in top and tail formation from Mountain Ash and Perhiwceiber to the phurnacite plant at Abercwmboi. This was the only time I managed to capture the dramatic departure and I used Ektachrome 200, which was a very grainy film at the best of times. Sadly, this transparency suffered from the poor E6 processing which blighted most chemists at the time and was covered in hundreds if not thousands of black blemishes. Because of the drama of the image, I have spent a couple of hours removing as many as I could in Photoshop! Hopefully it was worth it. **Photographic Details: Canon AE-1, Ektachrome 200, Tamron 200mm f3.5 lens, 1/250 at f5.6.**

Above: One problem with the Tamron interchangeable mount system was that more complex mounts like the Canon FD had a tendency to malfunction and fail to shut the lens down. This was the case on 12 March 1979. Looking at my notebook for the day, it seems I cycled to Cambridge's Mills Road bridge during the lunch break. Medical students in those days had lectures every day at 09.00 and 12.00 (including on Saturdays) and practicals in between these and during the afternoons. I arrived at the bridge at 13.15 and it looks like I stayed late, until just after 1400, because I was waiting for 8J21, the 12.20 Whitemoor to Temple Mills mixed freight, which didn't arrive until 13.57. It was a glorious day and unusually, the freight was hauled by 40124. I took four shots of the train and all were drastically over-exposed because the Tamron 200mm lens attached to my Canon AE-1 had failed to shut down. Hence my 1/250 at f8 was in fact 1/250 at f3.5. The transparency has languished in a folder for four decades, until I decided to see if it could be salvaged in Photoshop. This is the result. **Photographic Details: Canon AE-1, Fujichrome R100, Tamron 200mm f3.5 lens, accidentally over-exposed by 2.5 stops at f3.5.**

The 1980s

1980

This was the year that my freight only exploration kicked off in earnest. I had been appointed as the Visits Secretary for the Cambridge University Railway Club (CURC) the previous autumn and had managed to arrange a December 1979 visit to Temple Mills Yard and Wagon Works in north-east London, where we witnessed the final weeks of hump shunting at the once busy yard. The second visit I planned was to Toton in March 1980 and by then there were rumblings that we seemed to concentrate too much on freight visits! I managed to arrange some branch line trips to Bathgate and Forfar, before I was ousted for being unbalanced in my choice of visits! I hope in retrospect that the membership appreciated the amazing opportunities that the CURC provided, to visit and photograph the last knockings of wagonload traffic in the UK.

The year was also dominated by the Woodhead route, the demise of which was imminent. I made three visits to it that year, the most memorable of which was in April 1980. A group of six of us from the CURC took the train from Cambridge to Mexborough, taking our bikes with us so that we could cycle to various locations between Wath and Wombwell Junction. Carrying bikes on trains was relatively easy in those days, as trains had carriages purely for carrying parcels and luggage. In the summer my father loaned me his VW camper van (the family car) to do a UK wide tour, filling in gaps for my "Freight Trains of British Rail" book. I did some work in Germany again, but this time as a clinical research assistant, which left very little time for photography. Paul Shannon and I had a couple of crazy train and bicycle days out, including one in September of 1980 where we walked the length of Tinsley yard in the yellow vests we had been given during the Temple Mills visit. The year was rounded off with another visit to the Woodhead route, this time to Penistone, so all in all 1980 was an exceptionally productive year.

Below: Over the New Year, my younger brother and I stayed with my grandparents in Great Harwood in Lancashire and from there had a day out at Guide Bridge and Dinting. My brother kept me entertained with his comedy gags and we saw half a dozen freight trains. On 4 January 1980, 76046 crosses Dinting viaduct with a Dewsnap to Tinsley mixed freight. **Photographic Details: Canon AE-1, Kodak Tri-X, Tamron 200mm f3.5 lens, 1/250 at f11.**

Above: On 15 April 1980, half a dozen of us left Cambridge at 05.50 on the Peterborough train. There we changed onto the 05.50 Kings Cross to Aberdeen, which was hauled by 55016 and averaged 97.6 mph between mileposts 84 and 102. Changing at Doncaster we arrived at Mexborough at 08.53 and headed off towards Wath on our bicycles for the last four-mile leg of the journey. On the way, we were distracted by several locations, including a very rickety footbridge between Manvers Coke Works and Wath Junction signal box, where this image was captured. 20059 & 20005 head south from Wath with a rake of empty BEV steel wagons. The signal box in the background is for Wath Central Junction and had 50 levers. It only survived another year, closing in May 1981. The rest of the day was taken up cycling to Wombwell Junction and back. Hauled by 55007 from Doncaster to Peterborough, we arrived back in Cambridge at 20.43, tired but happy! **Photographic Details: Canon AE-1, Kodak Ektachrome 200, Canon 100mm f2.8 lens, 1/250 at f8.**

Top left: My first visit to Shrewsbury for photography was on 30 September 1980. I suspect having a girlfriend who lived in Lydbury North might have had something to do with the visit! That day saw 56033 & 56038 heading south past Crewe Junction with a special working, clearing the iron ore from the closed Shotton steelworks and moving this to Llanwern. **Photographic Details: Canon AT-1, Ilford FP4, Canon 100mm f2.8 lens, 1/250 at f8.**

Left: Perhaps my favourite image of the Woodhead route was taken on 29 October 1980. A day out on the train from Cambridge to Penistone allowed me four hours on the station, during which time I visited Huddersfield Junction signal box, wandered to the goods yard west of the station and managed to photograph nine Class 76 hauled freights. 76021 & 76008 pass Huddersfield Junction box with 8M17, the Wath to Ashburys, which is carrying household coal from Grimethorpe Colliery. **Photographic Details: Canon AE-1, Ilford FP4, Tamron 200mm f3.5 lens, 1/250 at f8.**

Above: On 2 June 1980, Paul Shannon and I made a visit to Barrow Hill. Our route there was circuitous, as we left Cambridge at 08.25, eventually arriving in Leicester at 11.16 where we hoped to photograph some freight passing the semaphore signals. We left Leicester at 12.31 and after changing at Derby, arrived in Chesterfield at 13.29. From here we cycled along the A619 to Barrow Hill where this image of 44004 heading a Toton to Tinsley mixed freight was captured. I was captivated by the location and visited several further times before the area was resignalled; the yards there form the basis for my model railway, Barrow Hill Junction. We were enjoying the location so much, that we suddenly realised we might not be able to get home and as I recall there was a frantic cycle to Worksop, which included a pause to photograph a Class 56 in Manton colliery. We were hauled by 55003 from Retford to Peterborough and got home at 22.55. **Photographic Details: Canon AE-1, Fujichrome R100, Canon 100mm f2.8 lens, 1/250 at f5.6.**

Above: A second image taken during my whistle stop tour of freight only locations, taken on 27 June 1980, is this image of 37016 leaving Consett Steel Works in Country Durham, with the 9P30 trip freight to Tyne Yard. There was optimism amongst the workforce at the time, because they were working incredibly hard to show British Steel how productive they could be and freight traffic was vibrant. The optimism was ill placed though, as just weeks later the plant closed, along with many others under the Villiers Review of outdated and inefficient plants. The workforce at British Steel was reduced from over 250,000 to just 130,000 as a result of this one review. The location of this picture is now a greenfield site and all memories of steel making in Consett are now fading. **Photographic Details: Canon AE-1, Kodachrome 64, Canon 100mm f2.8 lens, 1/250 at f5.6.**

Left: The June trip to fill in gaps for my first freight trains book allowed me to visit several off the beaten track Scottish locations. On 28 June 1980, a central Scotland shed bash included visits to Dunfermline Townhill shed and Thornton Yard, which was one of the more difficult locations to reach on public transport. In this view of Townhill depot are 08425 & 25011 and from left to right the Class 20s are 20203, 20227, 20221, 20224, 20201, 20222, 20223, 20217, 20216, 20191 & 20206. **Photographic Details: Canon AE-1, Kodachrome 64, Canon 100mm f2.8 lens, 1/250 at f5.6.**

Above: At the end of the day wandering from Shepcote Lane to Treeton Junctions on 23 September 1980, we were treated to the appearance of 40122 on the 6Z68 Lackenby to Corby extra steel coil train. My notebook reveals that I caught the 07.40 Cardiff to Newcastle to Sheffield and took the bus to Tinsley. It seems Paul Shannon and I walked from the bus stop, through Tinsley Yard to Shepcote Lane, then back along the yard and all the way to Treeton Junction. We then walked to Rotherwood and Orgreaves, before catching the bus back to Sheffield, where I caught the 18.53 back to Cardiff. I can't imagine what health and safety staff would make of such actions today! **Photographic Details: Canon AE-1, Kodachrome 64, Canon 50mm f1.8 lens, 1/250 at f5.6.**

1981

During 1981 I had the small matter of having to get my degree in Pharmacology, but in photographic terms, perhaps the most momentous event of the year was the introduction of Ilford XP1 film. Based on a colour emulsion, this promised medium format quality on 35mm film and was rated at 400 ASA. The year saw the last of my contributions to the CURC, with brake van trips in Scotland during March and some epic days out, the longest of which is recounted in one of the captions below. My final term in Cambridge was characterised by revision, during which I made several visits each day to catch particular freights. This was possible because a contact in Cambridge Yard freight office equipped me with a copy of the working timetable. The summer after graduation was relatively uneventful and I started clinical medicine in Oxford in September 1981. Unlike pre-clinical studies, there were no long summer holidays and as a result, photography was largely confined to Oxford and the neighbouring area.

Above: This is one of my favourite images of the Woodhead route, taken at Kendall Green Crossing on 4 March 1981. 76007 & 76026 head the 6M55 Wath to Fiddlers Ferry MGR train, which is being banked by 76006 & 76024. The itinerary for the day reminds me how enthusiastic we were back then! Paul Shannon and I left Cambridge at 01.14, hauled by 31200 and arrived at Peterborough at 03.01. We then travelled from Peterborough to Leeds behind 31126, leaving Peterborough at 03.08 and arriving in Leeds at 05.52. The 06.30 to Manchester was then taken and we alighted at Huddersfield at 07.02, in time to catch the 07.28 to Penistone. We had our bicycles with us and cycled from Penistone to Wombwell Junction and down the road that runs beside the Worsborough incline, photographing along the way. The next entry in the log shows that we boarded the bankers at Wombwell Junction at 17.00, leaving at 17.06 and rode in 76006 & 76024 as they pushed 76032 & 76034 on the 6M57 Wath to Fiddlers Ferry MGR service. We left our bicycles, care of the signalman at Wombwell Junction, returning at 17.55 to reclaim them. We then cycled to Mexborough, where we boarded the 18.55 to Doncaster, allowing us to catch a York to Kings Cross service hauled by 55018, getting to Peterborough at 21.56. We managed to make the 21.58 to Cambridge, getting home at 22.58. Now that is serious railway photography! **Photographic Details: Canon AE-1, Ilford FP4, Canon 100mm f2.8 lens, 1/250 at f5.6.**

Above: A second image from the epic 4 March trip of 20029 & 20031 winding out of Barrow Colliery, with the afternoon trip to the nearby Wath Yard. As I recall, we were on our bikes hurtling down the hill from Lewden Crossing when we spotted a pair of Class 20s high up on the hillside on the other side of the valley. We had no idea where the railway line was, but decided to investigate. The resulting pictures proved valuable, as the colliery and the branch line that served it closed not long afterwards.
Photographic Details: Canon AE-1, Ilford FP4, Canon 100mm f2.8 lens, 1/250 at f4.5.

Top right: Film number 206 was my first roll of Ilford XP1, which had been released in January 1981. This image, taken at Dundee West, was one that convinced me to switch over to XP1 for all my photographs. The lack of grain and the 400 ASA film speed opened up innumerable new possibilities in poor weather and allowed, with care, prints that looked like medium format to be produced from a 35mm camera. 40002 leaves Dundee after shunting in the goods yard with 7A18, a Thornton Yard to Aberdeen mixed freight. One other aspect of a long-forgotten world was the fact that I was travelling on a "Priv" or privilege ticket. There was a system in the late 1970s and early 1980s whereby each University college had a "Rail Rep" who had to field enquiries from students, thus reducing the workload at Cambridge station. The three Rail Reps who had fielded the most enquiries each term were given a BR privilege ticket. I was fortunate enough to be a Rail Rep for one year, in my final year (it was a coveted post and one had to wait until one's final year). I used this to acquire privilege tickets from Cambridge to Wick, Cambridge to Thurso and Cambridge to Penzance for the three terms I held the post, always making sure I had more signed for queries than anybody else in the university. **Photographic Details: Canon AE-1, Ilford XP1, Tamron 200mm f3.5 lens, 1/250 at f8.**

Below: Revision for my finals was interspersed with trips out on my bike to catch the Chesterton sand train at 08.30, the 8J21 Whitemoor to Temple Mills at lunch time and 8J93, the 16.20 Whitemoor to Temple Mills at tea time. Here, 37088 crosses the River Cam, just south of Chesterton Junction with 8J93 on 11 May 1981. Information from my contact in the freight yard at Cambridge revealed that the train weighed 1242 tons and contained 45 loaded wagons and 12 empties. As well as a ferry van and numerous loaded coal wagons, the consist contained 18 KRV steel carriers with steel for reinforcing concrete that was travelling from Scunthorpe Appleby to Rochester in Kent. **Photographic Details: Canon AE-1, Ilford XP1, Tamron 200mm f3.5 lens, 1/500 at f8.**

Above: A favourite freight location when I was studying in Oxford was the Morris Cowley car factory. Here, on 8 December 1981, 31416 heads along the branch from the facility and into the sunset with 4Z40, an additional service from Morris Cowley to Harwich with cars for export. **Photographic Details: Canon AE-1, Kodachrome 64, Canon 100mm f2.8 lens, 1/250 at f4.5.**

Below: The winter of 1981/82 was one of the coldest on record in the UK and I have since established that the coldest temperature ever recorded in Oxford was minus 16.6 degrees Celsius in January 1982. The cold spell started early however and this view taken on 12 December 1981 shows 50031 "Hood" accelerating north from Oxford past the bridge at Walton Well Road, with what is almost certainly the 13.50 Paddington–Liverpool, which it worked as far as Birmingham New Street. It was cold! **Photographic Details: Canon AE-1, Kodachrome 64, Canon 100mm f2.8 lens, 1/250 at f4.**

1982

The camaraderie of Cambridge was lost and the pressure of clinical medicine replaced it. Most of my photography was around Oxford, with the bonus that I was given my first car by my father, to facilitate attendance at clinical attachments. Little did he know (or maybe he did), that its major use was to get out and about taking railway photographs. A second driver to my photography during the year was the need to complete my second book with OPC – Diesels in South Wales, which had been commissioned with a handshake at a dinner with Rex Kennedy, who lived nearby in Wheatley, Oxfordshire.

Above: Apart from local outings, I undertook very little photography during 1982, except when I had holidays, which were all spent at my parents' house in South Wales. This wasn't just because I had nowhere else to go, but also because I was collecting material for my new project – the book "Diesels in South Wales". Here, on 15 June 1982, we see 37162 winding out of the docks in Swansea with 9B80, from Swansea to Pantyffynnon. There is no railway here anymore. **Photographic Details: Canon AT-1, Ilford XP1, Canon 135mm f2.8 lens, 1/250 at f5.6.**

Above: I have always thought this image was perfect for the railway modeller. 03141 races back to Llandeilo Junction Yard with 9B36, the Mondays, Wednesdays & Fridays only Llandovery to Llandeilo Junction freight. The train is captured at Rhosmaen and I chased it in my decrepit Renault 7 on 18 June 1982. **Photographic Details: Canon AE-1, Ilford XP1, Canon 100mm f2.8 lens, 1/250 at f5.6.**

Left: There is nothing exceptional about this image except that I enjoyed the view both ways from the overbridge at Somerton Road in Newport, which was especially dramatic looking east towards Llanwern. On 23 September 1982, 31286 heads west in perfect light hauling a Bristol Temple Meads to Cardiff service. **Photographic Details: Canon AE-1, Ilford XP1, Tamron 200mm f3.5 lens, 1/500 at f8.**

Left: This view, taken at Preston station on 10 December 1982 is of 81020, which has paused with 6V93, the daytime Mossend to Severn Tunnel Junction Speedlink working. Trying to decipher my notes reveals that the consist included two BBA wagons from Mossend to Ebbw Vale, two cement tanks from Eastgate to Radyr for repair, two OCA wagons to RNAD Trecwn, two cargo wagons from Law Junction to Bridgewater, a VBB with potatoes from Kirkcaldy to Truro, an FXV ferry van from Aberdeen to Eastleigh and two tanks with Nitric acid from Law Junction to Bridgewater. All of this is traffic that has since been lost to rail. **Photographic Details: Canon AE-1, Ilford XP1, Canon 100mm f2.8 lens, tripod, 1s at f16.**

1983

The year was predominantly taken up with studying and hence photography was mainly in and around Oxford or South Wales when visiting my parents during the holidays. One notable trip was in October 1983, when I spent a week in the North East as a guest of David Allen in Durham and explored many new freight locations.

Above: The significance of this location could never have been guessed back in 1983. For me it was a wonderful backwater with two or three trains a day and a signalman who enjoyed reminiscing. Amazingly he came from Padbury in Buckinghamshire, where my aforementioned first girlfriend had grown up; he remembered her father Wilf who was a railway enthusiast and was amused to meet her latest ex-boyfriend! Of course, this will now be the junction between the new East/West route to link Cambridge and Oxford (and what a mistake it was to close the original route) and the new HS2 route. Here we see the secondman of 25086 taking the token for the single line to Aylesbury with 6B04, the Northampton to Aylesbury parcels service. **Photographic Details: Yashica Mat 124G, Ilford XP1, 1/500 at f11.**

Right: On 26 February 1983, signalman Price strikes a pose in Heath Junction signal box, which is two miles north of Cardiff Queen Street. Heath was the nearest box to our family home on Nant Fawr Road and the first signal box I ever visited. It still had gas lamps, right until its closure in November 1984. Originally built with a 77-lever frame, it was much reduced by the time of this photograph, with just 25 levers left on the frame, this the result of the closure of the sidings at Heath Junction in the mid-1960s. **Photographic Details: Yashica Mat 124G, Ilford XP1, 0.5s at f8, resting on a chair.**

Below: Although I studied in Oxford for three years and lived in Green College, which was just a stone's throw away from the railway, I only managed to catch the Rewley Road coal depot trip working once. On a dismal day in March 1983, 08804 waits for the crew and the supervisor to wind the swing bridge across, before it propels four 21.5-ton MCV wagons, laden with coal from Abercwmboi, into the coal depot. The route to Oxford had been via one of the daily trip workings from Abercwmboi to Radyr yard, then it joined the evening mixed freight from Radyr to Severn Tunnel Junction. From there an overnight train took the wagons to Acton yard, from where they were tripped to Oxford Goods Yard on the daily mixed freight. The Class 08 had brought the wagons through Oxford station and is now propelling them for the last few yards of their lengthy journey. **Photographic Details: Canon AT-1, Ilford XP1, Canon 100mm f2.8 lens, 1/125 at f4.**

Above: A favourite line which many have photographed over the years was the Bury Port and Gwendraeth Valley route from Cwmmawr to Burry Port. Prior to capturing this image, I had made several visits to photograph the line and ridden in the guard's van along the route. The section of line between Burry Port and Kidwelly closed on 19 September 1983, so this image taken on 15 September 1983 was very much a last-minute attempt to record this stretch. From left to right, 03120, 03151 & 03144 trundle into the yard at Burry Port with a load of coal from Cwmmawr Washery. The last train from Cwmmawr was in 1996 and the line closed in April of that year. **Photographic Details: Canon AE-1, Kodachrome 64, Canon FD 70–200 f4 lens, 1/250 at f5.6.**

Below: On 14 October 1983, 37033 is seen running round its train at Seabanks on the Durham Coast line. My notebook at the time records the lens and exposure details and has the comment "v windy ? blurred". I had obviously been hit by a gust of wind off the North Sea. Fortunately the image is sharp and represents my first trip with a car exploring the County Durham mines. **Photographic Details: Canon AE-1, Kodachrome 64, Canon 100mm f2.8 lens, 1/250 at f5.6.**

1984

To say 1984 was eventful might be an understatement. I got married, graduated as a doctor and spent two weeks in a hospital in Kathmandu nearly dying from influenza. In order of occurrence, I went in March of 1984 to India, from where I had planned to travel overland to undertake my medical elective at a mission hospital in Amp Pipal in the Himalayas. Unbeknown to me, I had contracted influenza B before leaving and within days became very unwell. To cut a long story short, I was repatriated to the UK weighing just seven stone, having spent some time unconscious in a mission hospital in Kathmandu. It is sobering to think that influenza can make one so unwell. Of course, it didn't make the BBC news or lead to a lockdown and so the public are not kept informed about the severity of flu! Marriage and then working as a house surgeon in Oxford's John Ratcliffe Hospital severely reduced my photographic output, but one or two snaps were taken on our honeymoon and when visiting friends and family during weekends.

Above: In the depths of Lucknow Shed a sacred cow poses next to a couple of steamers. This was one of very few images that I managed to capture on my overland journey from Delhi to Pokhara via Lucknow and Gorakpur. Just after I took this picture, the shed master came to tell me I was not allowed to take pictures, but my guide (a patient from the nearby missionary hospital) could, because he was Indian. Sadly, the patient was a schizophrenic whose medication gave him a terrible intention tremor and consequently, the rest of the day's pictures came back badly blurred! **Photographic Details: Canon AE-1, Ilford XP1, Canon 50mm f1.8 lens, 1/15 at f4.**

Top right: On 17 September 1984, during a week of annual leave, I made several short day trips from Oxford to the South Midlands, including this visit to Hawksbury Lane, where I was lucky enough to catch 47223 shunting. After what seemed like an interminable wait, it left with 7E88, the Bedworth Murco to Humber Oil Refinery empty tanks and is captured passing the signal box.
Photographic Details: Canon AT-1, Ilford XP1, Canon 100mm f2.8 lens, 1/250 at f8.

Right: In July 1984, the miners' strike was in full swing and I found myself driving from Lichfield to Lincoln while on our honeymoon. As everybody knows, Mansfield Concentration Sidings are directly en-route! With the Nottinghamshire coalfield being just about the only one functioning, it was possible to see a coal train; the nearby Shirebrook Shed was packed with silent locomotives even though it was mid-week. 56111 backs into a siding at Mansfield Concentration Sidings, with the 6T05 trip working from Toton. The yard behind had been severely rationalised, compared to what I found on my first visit in 1980. 56121 is just visible and is heading light engine back to Shirebrook Depot. **Photographic Details: Canon AE-1, Kodachrome64, Canon 100mm f2.8 lens, 2/250 at f5.6.**

HAWKESBURY LANE

56111

Above: Sunset at Arnside on 19 July 1984 as 40035 trundles over the viaduct with the evening chemical train from Corkickle to Ince and Elton. Shortly after this shot was taken, I slipped and landed in the muddy estuary, ending up coated with mud from head to toe! **Photographic Details: Canon AE-1, Kodachrome 64, Canon 100mm lens, 1/250 at f4–5.6.**

Below: A fleeting visit to my college friend Paul Shannon on 19 September 1984 saw outings to Wigan, Warrington and Northwich. The driver of 20166 & 20174 makes his weary way to the mess room at the end of his shift in the freight loops to the east of Northwich station. The train is the 6F42, which he had driven from Northwich to Tunstead and then back to Northwich. Another driver will complete the final leg of the train's journey to the nearby Oakleigh Sidings. **Photographic Details: Canon AE-1, Kodachrome 64, Canon 50mm f1.8 lens, 1/250 at f5.6.**

1985

For the first half of 1985, I worked as a house physician in Worcester Ronkswood Hospital and the job had the bonus of a fixed afternoon off each week. This allowed some exploration of the Worcester area and as the days grew longer, of the West Midlands and as far as Toton and Leicester. The second half of the year was taken up with a new job teaching anatomy in Newcastle University, a prerequisite to a career in surgery. This opened up a whole new area for me to explore, which at the time was riddled with collieries and freight only branches.

Left: On 2 May 1985, it seems I headed out after work to catch the 6V70 Speedlink at Droitwich, before retiring to the pub next to Dunhampstead Crossing to enjoy a pint of cider and a bite to eat. It was light until well after 20.00 and the signalman reluctantly allowed me into the box to capture this view. I loved the carpet tied to his shoes to avoid dirtying the box floor and his far away gaze somehow captures a moment in time that was rapidly vanishing from our railway, as signal boxes disappeared at a rate of knots. **Photographic Details: Canon AT-1, Ilford XP1, Canon 28mm f2.8 lens, 1/60 at f8.**

Below: One of my favourite freight shots is this view at Droitwich, taken on 30 May 1985, of 45020 heading 6V70, the Cliffe Vale to Severn Tunnel Junction Speedlink. The train is made up entirely of Tiger Leasing china clay wagons and is returning from the Potteries to St. Blazey. With an excellent model of these wagons made for OO gauge and the recently released Bachman Class 45 with marker lights, it is possible to run an exact replica of this train on Barrow Hill Junction, my model railway! 6V70 was conveniently timetabled to pass Droitwich and then Worcester between 18.00 and 19.00, meaning it could usually be photographed after work on days when I was not on-call. **Photographic Details: Canon AT-1, Ilford XP1, Canon 300mm f5.6 lens, 1/250 at f11.**

Below: Having only scratched the surface of the Blyth and Tyne network during my 1980 visit, moving to Newcastle allowed the exploration to start in earnest. One of my favourite spots was a mound of earth adjacent to the flat crossing, just north of Ashington signal box, from where this view was taken. Friends often joked that my 300mm lens was glued to the front of my camera and this trusty lens was used here to capture 56132 hauling a Butterwell Colliery to Blyth Power Station MGR service past the NCB signal box at Ashington Loop on 15 August 1985. **Photographic Details: Canon AT-1, Ilford XP1, Canon 300mm f5.6 lens, 1/250 at f8.**

Right: Very near the end of the life of the Philadelphia Colliery system, on 28 October 1985, I managed to catch a train leaving the exchange sidings with coking coal from the Lambton Coking Works, hauled by 37173. Just visible between the wagons in the background is NCB shunter 513 (ex-BR Class 11, number 12098). For me, the man and his dog capture something of an age that was rapidly disappearing. The train is, 9J91, a Lambton to Tyne Yard trip working and the coal is bound for Scunthorpe Steelworks. The coke works closed in early 1986, not long after this photo was taken. **Photographic Details: Canon AT-1, Ilford XP1, Canon 50mm f1.8 lens, 1/125 at F5.6.**

Below: On a bitterly cold 12 February 1985, HST power car 43164 is captured on the bridge over the Worcester Canal, between Worcester's Shrub Hill and Foregate Street stations. My notes suggest this was a Great Malvern to Paddington express, but I'm not sure whether that is correct or this was simply a Hereford to Paddington train. **Photographic Details- Canon AE-1, Kodachrome 64, Canon 100mm f2.8 lens, 1/500 at f5.6.**

Above: Dudley Freightliner Terminal was one of the first container depots to open under the plans that Beeching drew up and also one of the most profitable; it closed in September 1989. On 30 April 1985, 45016 passes the depot during a fortuitous break in the cloud cover with 6V70, the Cliffe Vale to Severn Tunnel Junction Speedlink working. **Photographic Details: Canon AE-1, Kodachrome 64, Canon 50mm f1.8 lens, 1/250 at f4–5.6, half a stop over-exposed.**

Top right: Another location I discovered in 1985 whilst exploring the East Midlands during an afternoon out from Worcester, was Bestwood Park to the north of Nottingham. 20019 & 20004 pass the sidings and signal box with 6T71, the afternoon trip working from Hucknall Colliery to Toton on 2 July 1985. My notes suggest I arrived just as the train appeared round the corner and that I took a black and white image with my 300mm lens and then a very rushed colour shot which I feared wouldn't come out – fortunately it did. **Photographic Details: Canon AE-1, Kodachrome 64, Canon 100mm lens, 1/250 at f5.6.**

Right: By 1985 I had written several articles for Rail Enthusiast and found that it was possible to arrange visits to many yards and depots, as long as the local railway management could be promised an article in the magazine, showcasing their operations. I also had the beginnings of an idea to try and write a reference book about Britain's rapidly disappearing marshalling yards, which was later published by OPC/Haynes in 1988. On 26 September 1985, I made an official visit to Tinsley to enable me to put together an article on the yard for Rail Enthusiast. Whilst chatting to staff at the west end of the complex, the driver of 20086 & 20150, Mr. L. Middleton, suggested that I might like to ride in the cab to Aldwarke and back on the 6T35 afternoon trip. As we passed Ickles Steelworks he suggested a stop with the cooling towers as a backdrop and that I hop out and take a picture, on the proviso that I sent him a copy. I still have his address in my notebook and duly sent an 8 x 10 inch print to him. The signalman at Rotherham Road couldn't understand why we had taken so long to reach him from Tinsley East Junction! **Photographic Details: Canon AE-1, Kodachrome 64, Canon 100mm f2.8 lens, 1/250 at f5.6.**

20086
HST

1986

1986 was another great year for photography, mainly because I unexpectedly passed the anatomy exams required to continue in surgery six months early and therefore spent six months teaching at Newcastle University, without the need to study each evening. The second half of the year was spent as a casualty officer in Newcastle General Hospital, in what was then known as the "Wild West End" of Newcastle. This involved working a week of 12-hour days, then a week of 12-hour nights, followed by a few days off, which were perfect for outings to Fife and Glasgow amongst other destinations.

Below: The wonderfully named Bedlam Crossing was just north of the Three Spires Coking Plant in Coventry. Here on 24 February 1986, 56063 heads north with 6D85, the Three Spires to Toton working. The signal box never had a name board until the end of its life when apparently somebody took a road sign from Bedlam Lane and propped it against the cabin.
Photographic Details: Canon AT-1, Ilford XP1, Canon 100mm f2.8 lens, 1/500 at f11.

Above: On 2 May 1986, I made one of several trips to Fife. My notes record that having arrived at Thornton Yard at 06.00, several drivers offered me rides on their trip freights. Driver Adam Hutchinson was especially keen that I joined him and his crew on the Auchmuty trip, but not until I had done the first trip of the day with driver Jack Storrar to Methil Docks. Driver Hutchinson ended up waiting 20 minutes for me at Markinch, so I could join him for the winding trip to Auchmuty Paper Mill, through what was the tightest squeeze on BR; this tunnel had less than an inch clearance on either side of 08515, which is seen approaching it with its load of a single Tullis PAA wagon from St. Blazey. The inset shows the crew at the end of the line at Glenrothes. **Photographic Details: Canon AT-1, Canon 50mm f1.8 lens. Exposure details not recorded.**

Above: On 24 October 1986, David Allen and I had a day out on the Redmire branch. We were offered a cab ride from the last level crossing to the loading point at Redmire and back, after which we chased the Redmire to Redcar train, as it made its way back along the branch to Leeming Bar. As the train had to stop to allow the guard to open and close various crossing gates along the route, it was easy to get three or four shots along the branch. 20306 & 20305 have just collected the token for the next section at Bedale, as the signalman watches the train pass. This was the first week that 20s had begun operating in the area, as six members of the class had just been allocated to the nearby Thornaby depot for the Redmire and Boulby workings. This later expanded to 12 locos. **Photographic Details: Canon AT-1, Ilford XP1, Canon 100mm lens, 1/125 at f5.6.**

Below: On 8 April 1986 I was allowed an official visit to Ravenscraig Steelworks and the four yards which supplied it. Awaiting departure from No.1 Yard is 20127, with the 6T12 trip working to Mossend, which is made up of MXV 16-ton mineral wagons loaded with scrap bound for Shieldmuir Steelworks. **Photographic Details: Canon AE-1, Kodachrome 64, Canon 100mm f2.8 lens, 1/60 at f5.6–8.**

Right: On 2 May 1986, 20227 has arrived at the distillery at Cameron Bridge with the second trip of the morning. It collects two CO^2 tanks bound for London which will travel via Thornton Yard. From there, they will be collected by the afternoon Perth to Mossend Speedlink train and then travel overnight on a Mossend to Willesden Speedlink service. Bachman have just produced an excellent retooled model of 20227, which is now preserved and at the time of writing in 2022, was operating at the North Norfolk Railway. **Photographic Details: Canon AE-1, Kodachrome 64, Canon 100mm f2.8, 1/250 at f5.6.**

Below: An image that always makes me chuckle, as the signalman at Ashington Loop salutes the passing No. 507 on 28 August 1986. The ex-BR Class 14, which was formerly numbered D9525, is hauling loaded internal user coal hoppers from Ashington's coal stocking site to the Ashington Coal Concentration Depot. This was a transparency that I was persuaded to submit for use in a book and which like several others, needed an hour's work to remedy the damage done when it was stuck to a drum scanner to allow reproduction. Even now I see blemishes every time I review the image! **Photographic Details: Canon AE-1, Kodachrome 64, Canon 50mm f1.8 lens, 1/250 at f5.6.**

1987

1987 was my first year as a surgical registrar which involved working a "1 in 2" (every other night and every other weekend) as well as the normal 07.30 to 17.30 working day – an average of 120 hours a week. I learnt an enormous amount, but was permanently knackered! Photography was restricted to the odd summer evening, occasional Saturdays in the North East and holidays, which again were mainly spent at my parents' house in Cardiff, although one holiday was taken in Cambridge visiting friends. Conveniently, the Welsh visits and the East Anglian jaunt allowed me to fill in some gaps for the Freight Only series of books, which were published between June 1987 and November 1988.

Above: Photographed arriving at Margam yard on 31 March 1987, 47099 heads 6A35, the Velindre to Margam trip working which is returning empty BBA and BAA wagons that had delivered rolled coil to the tin plate factory at Velindre. In the distance is Margam Abbey Works East signal box which closed on 1 November 1987. The larger Margam Sorting Sidings box, which is behind the photographer, closed on the same date. Also visible in this shot is 37304 leaving Margam with 6Z68, a Saltend to Baglan Bay freight, the Z headcode denoting a special working, due to the fact that it had travelled to Tidal Sidings "in error" the day before! 08897 can be seen waiting to return from the steelworks to Margam yard. Virtually nothing is left of this thriving scene and the vast expanse of Margam Yard has returned to nature. **Photographic Details: Canon AT-1, Ilford XP1, Canon 100mm f2.8 lens, 1/250 at f4-5.6.**

Above: Because days out taking pictures were determined by my holidays, many of my photographs have been taken in bad weather! In pouring rain on 2 April 1987, 08769 trundles past the platform at Pantyffynnon with a trip working from the yard to Wernos Washery on the other side of the Central Wales line. On the right, 37241 can be seen in charge of 6C45, a Speedlink coal working that began at Coedbach and which will continue on to Exmouth Junction. The loaded HEA wagons are carrying coal that is destined for Filton, Neasden, Wapping and Stonehouse. **Photographic Details: Canon T-1, Ilford XP1, Canon 50mm f1.8 lens, 1/125 at f4.**

Below: A slight let up in the rain on 2 April 1987 sees a Class 37 paused at Ammanford Crossing with coal from the nearby Bettws Drift Mine for washing at Wernos. Shortly after this shot, the heavens opened again and a driver took pity on me, offering a ride to Gwaun-cae-Gurwen and back which I gratefully accepted. **Photographic Details: Canon AT-1, Ilford XP1, Canon 300mm f5.6 lens, 1/250 at f11–16.**

Above: With every other weekend off from 17.00 on Friday until 07.30 the following Monday, much time was spent sleeping and eating, but just sometimes there was enough energy left for a Saturday afternoon drive to somewhere local. One favourite spot was Sunderland Docks and the depot there, which would host half a dozen or more of the locomotives used to haul Durham coal. On 19 April 1987, 56119, 56129, 56128, 56131, 56112, 56123 & 56122 line up for a sunny snap along with yard pilot 08442.
Photographic Details: Canon AT-1, Ilford XP1, Canon 100mm f2.8 lens, 1/250 at f11.

Top right: This picture reminds me of an epic day out on 15 July 1987. The late Ken Groundwater was helping Peter Robinson with a past and present book about Teesside. Ken worked as a supervisor at Tyne Yard and had arranged a day out, which included a ride along the branch line that ran from Port Clarence to Seal Sands on the northern banks of the River Tees. He had asked if I would like to join him and a great day out was had by all, when we managed to photograph some rare corners of the network. Here, 20076 has just arrived at Port Clarence with the 6P01 trip working from Tees Yard. It has brought two TTA tanks of acetic acid for storage in Seal Sands, from where the train was supposed to pick up four empty sulphuric acid tanks, but these weren't ready, so we returned light engine. **Photographic Details: Canon AT-1, Ilford XP1, Canon50mm f1.8 lens, 1/125 at f11.**

Right: Remarkably in 1987 and 1988 two different unfitted freights still ran on the former Blyth and Tyne railway network. Here, the 6B10 trip working from Lynemouth Colliery to Blyth Staithes, made up of eight HUO unfitted hoppers, leaves the colliery behind 56013 on 16 July 1987. Behind it, 56118 will follow with the 6B09 from Lynemouth to Blyth Cambois Power Station. The frequency of these MGR services had just increased to three per eight-hour shift, from the previous year when just one ran during each shift. One other reason for including this image was that it took centre stage in an exhibition in the window of the Halifax Building Society on Gosforth High Street in Newcastle. As I saved my royalties from photographs at this branch (to go towards the 1992 China trip), the staff there became interested in the pictures I was taking and held a display of the Blyth & Tyne Railway for some months.
Photographic Details: Canon AT-1, Ilford XP1, Canon 100mm f2.8 lens, 1/500 at f11.

30

Ely
North Junction
25
ENJ
19

093

Above left: Ely North Junction has always been a favourite location of mine, ever since I first visited in February 1979, when several of us from the CURC spent the afternoon in the signal box there. On 9 September 1987, 37215 passes with 6N84, the daily Speedlink working from Duxford to Tyne Yard. Trying to decipher my notes, it seems that this left Duxford with glue bound for Stirling and Hexham for use in making chipboard, as well as empty methanol tanks and empty urea tanks both of which were destined for Billingham. During its stop in Cambridge Yard, grain from Newmarket was added to the front, as well as empty molasses tanks returning from Ely Papworth to Menstrie in Fife. The signal box at Ely North Junction opened in 1926 with a 76-lever frame and closed on 27 April 1992. **Photographic Details: Canon AT-1, Ilford XP1, Canon 200mm f4 lens, 1/250 at f8.**

Left: On 5 July 1987, this classic view of Toton depot was captured from the A52 embankment, albeit with a 300mm lens. We were on our way from Newcastle to South Wales and Toton provided the opportunity for a break in the journey. **Photographic Details: Canon AE-1, Kodachrome 64, Canon 300mm f5.6 lens, 1/125 at f5.6–8.**

Above: The iron ore trains between Port Talbot Docks and Llanwern have been variously hauled by triple Class 37s, pairs of Class 56s, Class 59s and then Class 60s until steelmaking ceased at Llanwern in 2001. Taken from the infamous footbridge adjacent to Cardiff Canton depot (now demolished) on 7 July 1987, 56031 "Merehead" & 56032 "Sir De Morganmwg County of South Glamorgan" are seen heading west with an empty iron ore working. 37805 is stabled on Canton along with several other Class 37s. **Photographic Details: Canon AE-1, Kodachrome 64, Canon100mm f2.8 lens, 1/250 at f5.6 (halfstop over).**

Above: It is hard to imagine that King's Lynn once had a thriving, rail connected commercial port. On 7 September 1987, 08757 shunts empty HAA hoppers onto the quayside to pick up imported coking coal for the cement works at Foxton. After delivering this to the yard at King's Lynn, the shunter returned to collect steel that had been imported from Germany.
Photographic Details: Canon AE-1, Kodachrome 64, Canon 50mm f1.8 lens, 1/250 at f5.6.

Left: On 8 July 1987, with Deep Navigation Colliery visible in the background, 37278 (near) and 37898 haul a rake of empty MDV 21-ton wagons from Nelson Shale Disposal Point. For several years, spoil from Taff Merthyr Colliery was dumped near Nelson and the short two-mile shuttle ran two or three times each day, with locomotives at either end of the train.
Photographic Details: Canon AE-1, Kodachrome 64, Canon 100mm f2.8 lens, 1/250 at f5.6.

1988

In 1988 I passed my final FRCS as well as undertaking registrar jobs in urology and general surgery in Ashington. When working in Ashington, my on-call room backed onto the main line south from Ashington and I had the dubious privilege of being woken during the night by Class 56s hauling coal south from the mines at Lynemouth and Butterwell. Being sleep deprived enough already, this wasn't actually welcome. My photography during the year was the usual mix of local outings in the North East and holiday trips to stay with my parents in Cardiff.

Right & below: I first visited the amazing Harton Electric Railway in 1985, just weeks after taking up my first post in Newcastle. Due to pressures at work, I didn't return there until 18 April 1988, when these two images were taken and rumours that the system would soon close were surfacing. Taken just two minutes apart, they show Bo-Bo electric locomotive No. 12 at either end of the tunnel which connected St. Hilda Colliery Sidings to the staithes on the River Tyne. Both show the driver leaning out of the cab window to check the photographer was well clear of the line. My notes reveal that two rolls of XP1 film were exposed between 15.00 and 17.30 that day and that I had several rides through the tunnel. Hurtling along at what seemed like breakneck speed (probably only 20 mph), was both terrifying and exhilarating, bursting out of the tunnel where the momentum of the train allowed it to ascend, before it reversed onto the staithes in order to dump shale into a waiting coastal ship. **Photographic Details: Canon AT-1, Ilford XP1, Canon 200mm f4 lens, 1/250 at f5.6 and Canon 28mm f2.8, 1/250 at f5.6.**

Above: Another photogenic spot within a 20-minute drive of my home was Bedlington station, which had lost its passengers services back in 1964. On 25 May 1988, 56132 has just run round the 6B09 trip working, an empty MGR train from Cambois Power Station, which is heading to the opencast disposal point at Widdrington. **Photographic Details: Canon AT-1, Ilford XP1, Canon 100mm f2.8 lens, 1/250 at f11.**

Right: I must have made over a dozen visits to Abercwmboi signal box from 1977 onwards and on 19 October 1988, I took shelter there and enjoyed a cup of tea from Terry the signalman. He told me that the box was due to close and the 1884 vintage 38-lever structure did indeed close on 9 October 1989; but not before I squeezed in a couple more visits. 37702 waits in the loop with 7Z84, an MGR train from Tower Colliery to Aberthaw Power Station, to allow the northbound passenger service to Aberdare to pass. **Photographic Details: Canon AT-1, Ilford XP1, Canon50mm f1.8 lens, 1/60 at f8.**

Left: One of the biggest shocks of the 1980s was the 1986 South Wales Freight Train Plan, which advocated the complete closure of Severn Tunnel Junction Yard. This took place in November 1987, when Speedlink traffic was diverted to Gloucester, East Usk and Cardiff Tidal Sidings, which for a time became quite busy. On 15 July 1988, 37015 eases out of the yard, shunting the wagons for the 6M83 Speedlink working to Bescot. The train had two POA scrap wagons bound for Leicester, two OAA open wagons which ran empty to Gloucester and two TUA chemical wagons being returned from Gower Chemicals in Swansea to Bescot. 08796 has just arrived alongside 37015, with a single VGA van, which had been tripped from the docks, where it had picked up imported timber. **Photographic Details: Canon AT-1, Kodachrome 64, Canon 100mm f2.8 lens, 1/15 at f5.6–8.**

1989

The year was spent completing my registrar training in Newcastle and moving into full time research towards an MD thesis. This meant freedom from regular nights on-call, but also irregular hours in the laboratory, especially when experiments didn't work out quite as expected, which often happened. One notable event during this research job was a trip to the Mayo Clinic in the United States, to learn the laboratory techniques that were essential for my research. It was during this visit that I discovered the American hump yard (Northtown Hump Yard in Minneapolis), a discovery that led to two best-selling railway books about North American railyards.

Above: Between August 1988 and the end of January 1989, I worked in Ashington. At lunchtime, or when the days were longer, I often popped down to Cambois to see what was going on. On 26 January 1989, coal was being loaded into the Elisabethe from North Blyth Staithes. The coal came from Butterwell and was bound for Esbjerg in Denmark. I suspect I must have been a familiar face around the place during those six months because I was allowed onto the staithe and up the lighting masts to capture this overall view of a now long-gone activity – exporting coal from England! **Photographic Details: Canon AT-1, Ilford XP2, Canon 50mm f1.4 lens, 1/250 at f8.**

Left: Taken with the Yashica Mat 124G, this image reflects the friendliness of many railwaymen during the 1970s and 1980s. I had been having a cup of tea in the signal box at Crag Hall when this Middlesbrough Goods Yard to Boulby empty potash train was offered to the signalman. I said I wanted to try and get a shot of the train passing the box but couldn't work out where to stand. He explained that the best view was up the home signal, which he had to climb regularly to fill the paraffin lamps. He suggested I climb up and get a shot from there, which I duly did. 20008 & 20122 have just arrived and the single line token is about to be handed over before they wind into the loop at Crag Hall on 30 January 1989. **Photographic Details: Yashica Mat 124G, Fuji Provia, 1/250 at f5.6.**

Left: I couldn't leave out this classic view from the embankment of the A52 overbridge to the north of Toton Yard. Hundreds of photographers must have stood there over the years and I was a regular visitor until the palisade fencing went up at the entrance to Toton depot. When working in Newcastle and visiting my parents in South Wales, Toton was the perfect spot to break the drive. I did this on 26 October 1989, when 58020 was captured heading north with an empty MGR train from Ratcliffe Power Station, whilst 56018 awaits its path north in the down loop. Partially obscured by the exhaust from 58020 is a Class 37, which looks to be hauling the Spondon to Saltend empty acetic acid tanks train. **Photographic Details: Canon AT-1, Ilford XP2, Canon 300mm f5.6 lens, 1/250 at f8.**

Left: During the electrification of the East Coast Main Line, there were many weeks when all traffic was diverted via the Leamside Line, which ran between Ferryhill and Pelaw Junction to the south of Newcastle. One such day was 17 March 1989, when 47422 was captured in perfect spring light, hauling 1E57, the 11.01 Liverpool–Newcastle over Victoria Viaduct and the River Wear. **Photographic Details: Canon AE-1, Kodachrome 64, Canon50mm f1.8 lens, 1/250 at f5.6.**

Bottom left: One of my very first films in the Mamiya 1000S was taken during an official visit to Knottingley as part of an article for Rail Enthusiast.. 56073 is nearest to the camera, in the new Trainload Coal livery. Sadly, I did not record all the locomotive numbers, but I selected this image because it reminds me of the time when, even after the miners' strike of 1984, coal was still king, both with regard to electricity generation and railway freight haulage. **Photographic Details: Mamiya 1000S, Fuji Vevia, Sekkor 80mm f2.8 lens, 1/250 at f5.6.**

Below: Looking back over the years, it is amazing how small incidents can lead to major projects and even lifelong passions. In 1989, my laboratory research for my MD thesis was not going well and it was decided that I should visit the Mayo Clinic for a month, in order to learn the techniques needed to complete my MD. While flying from London to Minneapolis with North West Orient, as they were then called, we had to wait to land in Minneapolis and circled over the city several times. We flew over the city's hump yard, which is known as Northtown Yard and at the time was operated by the railroad company Burlington Northern. Bear in mind that this was 1989, when the UK's only operational hump yard was at Scunthorpe West, so seeing this massive yard stretched out beneath me was a revelation. My hosts were family friends from our time living in Rochester Minnesota during the 1960s and they kindly offered me an ancient Oldsmobile, with which I could explore at weekends. I asked if they would mind me making the trek to the Twin Cities (Minneapolis and Saint Paul) one Saturday and with their permission off I went. Having found the yard, I was welcomed into the control tower from where this image was taken. I asked the yardmaster about how many other hump yards there were in the USA and he said he had no idea, but there were at least ten on Burlington Northern's network. Thus, a fascination was born – where are these yards and how may were there in the USA? The end result of what I first saw above Minneapolis and this image were two books which sold tens of thousands of copies – North American Railyards of 2003 and 2014, which I hope to reprise in 2023. This was my first photograph of a hump marshalling yard in North America. **Photographic details: Canon AT-1, Kodachrome 64, Canon 50mm f1.8 lens, 1/250 at f 5.6.**

The 1990s

1990

All of this year was spent in the laboratory undertaking research into gallstones, CCK receptors and mucous! The advantage of laboratory work was the facility to work all weekend and then sometimes have a day off during the week. This allowed me to have some great days out to Humberside and Fife, and holidays continued to be spent with my parents in South Wales, who loved seeing their two grandchildren.

Above: With East Usk Yard still being used as the main Speedlink yard for South Wales, I made acquaintances with several staff there, which led to me being allowed to climb the lighting masts on a semi-official basis. This view, which was captured on 30 March 1990, is one of my favourites and shows 37799 with 7B71, an empty MGR train from Llanwern to Margam. When up the light mast, the backdrop of Llanwern Steelworks was truly spectacular. **Photographic Details: Canon AT-1, Ilford XP1, Canon 300mm f5.6 lens, 1/500 at f11.**

Right: This third steelworks-themed image from 1990 shows Scunthorpe Steelworks on 21 March, which at the time of writing in 2021, is the only one of the three steel-making plants that still operates. 08543 arrives at Dragonby Yard (formerly the reception for Normanby Park Steelworks which closed in 1980) with a trip working from Scunthorpe Trent Junction, carrying steel bound for Flixborough Wharf, while 08632 waits to take the train to Flixborough Wharf. The picture was taken half-way up one of the derelict lighting masts, in order to include the four blast furnaces at Scunthorpe in the background and involved swinging round some barbed wire at about 20 feet off the ground! I was allowed to ride with the train to Flixborough, returning in the guard's van by prior arrangement and this was my favourite image from the day. **Photographic Details: Mamiya 1000S, Ilford XPi, Sekkor 80mm f2.8 lens, 1/500 at f8.**

Above: The daily Lackenby to Corby steel coil working is seen here passing South Bank in Middlesborough on 25 November 1990. From memory, this was one of the first times the train had a Class 60 on the front, which at the time was a disappointment as I was hoping for the regular pair of Class 37s. Time has tempered that view. **Photographic Details: Mamiya 1000S, Fuji Provia, Sekkor 80mm f2.8 lens, 1/250 at f.5.6–8.**

Above: I first started taking pictures at Radyr Yard in 1977, when I used to cycle from Cyncoed in Cardiff, to have "tea" with a young lady. Those were the days. The yard was very busy back then, with permissive block on the approaches from Walnut Tree Junction and at "tea time" several coal trains would stack up on the freight lines from the north. Traffic dwindled, but saw a brief upturn in January 1989, when it was designated a centre for Speedlink Coal traffic. The renaissance was short lived though, and the last train left the yard on 19 June 1993. This entire scene is now covered with a housing estate, through which runs the Cardiff to Radyr line via Radyr Quarry Junction. On 30 March 1990, 37146 starts up with what will be the 7Z84 extra to Alexandra Dock Junction Yard, whilst 08654 hustles wagons out of the way to allow the departure. And yes, a lighting mast was climbed to get a good overall view! **Photographic Details: Mamiya 1000s, Fuji Provia, Sekkor 80mm f2.8 lens, 1/250 at f8.**

Bottom left: During 1990, I realised my photograph collection had a few gaps, in the form of the freight only lines in Nottinghamshire and Derbyshire, so I had a couple of long days out from Newcastle to address this. On 4 April 1990, 56023 passes Welbeck Colliery Junction, which as its name suggests is where the branch line to Welbeck Colliery began, with the 6Z38 working from Cottam Power Station to Ollerton Colliery. Standing in the loop is sister engine 56020, which has come down from Welbeck colliery to stable, allowing the driver to take his PNB (personal needs break). **Photographic Details: Mamiya 1000S, Fuji Provia, Sekkor 150mm f3.5 lens, 1/250 at f8.**

Right: A classic location, which is now obscured by trees, is the curve at Topley Pike, to the south of Peak Forest. During a disastrous holiday in Buxton with friends (the disasters were all my own doing, as I kept getting offered cab rides and signal box visits much to the disquiet of our lovely friends), I ventured out to this location. 47599 rounds the curve with the 6T82 trip working from Peak Forest to Hindlow. I still haven't worked out how to combine railway photography with family holidays, although as later images will show, it has occasionally worked! **Photographic Details: Mamiya 1000s, Fuji Provia, Sekkor 150mm f3.5 lens, 1/250 at F8.**

Below: From the days of Speedlink and Railfreight, this is perhaps my most cherished image. I had visited Fife from Newcastle half a dozen times since 1986 and had always missed the trip working to the Royal Navy Armament Depot (RNAD) at Crombie. Once when it ran, I got lost in the lanes leading to the branch line. On two other occasions, after leaving Newcastle at 03.00, I arrived at Thornton Yard to find it had been cancelled. On 17 August 1990, all the stars aligned and I rode the trip from Thornton Yard to Crombie, then on to Townhill where we ran round and visited Rosyth Dockyard RNAD. Driver and guard, William Arnott and J. Finnie, took me into both RNAD depots with instructions to stay in the locomotive's cab and just look like I belonged there. 20137 & 20118 reverse into RNAD Crombie with two VEA vans containing explosives for the depot. **Photographic Details: Mamiya 1000S, Fuji Provia, Sekkor 150mm f3.5 lens, 1/250 at f8.**

1991

During the first half of 1991, I finished the research for my MD thesis and spent much of the summer writing it up for submission to Newcastle University. In September of that year, I left the North East to take up a registrar job in Bristol, returning to on-call duties on a 1 in 4 basis (72 hours a week). Frequent visits across the Bristol Channel meant that again much of my photography was undertaken in South Wales. One notable trip was to present research work at the Digestive Diseases Week in New Orleans, which provided me with a second opportunity to photograph North American railroads, including from the inside of several signal towers in the city. I didn't however find any more hump yards on this visit.

Above: Surprisingly, as I had lived in Cardiff since 1961 and revisited the city to see my parents frequently since 1978, I had never ventured deep into the dockland area of the city. On 2 April 1991, I made my first visit to the end of the line in the docks, riding on a trip freight from the Allied Steel and Wire Castle Works to an export storage warehouse, where its load of steel reinforcing wire would be stored while awaiting onward shipment. The route pictured here was to be closed just a few months later as part of the redevelopment of the dock area into a marina and luxury accommodation. **Photographic Details: Canon AT-1, Ilford XP1, Canon 100mm f2.8 lens, 1/250 at f5.6.**

Below: My trips between Newcastle and South Wales alternated between the route via Toton and a more westerly route via Carlisle and the M6, on which Warrington Arpley Yard was the perfect spot for a break! I have included this image quite simply because I have always liked the curve and tilt of the train. On 11 April 1991, 20055 & 20028 wind out of Arpley Yard with 6T76, the Fiddlers Ferry Power Station to Bickershaw Colliery empty MGRs. It had run round at Latchford Sidings and again at Warrington Walton Old Yard, before heading north along the West Coast Main Line and branching off to Bickershaw. **Photographic Details: Canon AT-1, Ilford XP1, Canon 300mm f5.6 lens, 1/500 at f8.**

Above: After my exploration of Cardiff Docks in April 1991, I decided to explore the dock areas at Barry, Swansea and Newport. This was of course in the days before palisade fencing and check points, and so it was easy to follow the railway lines down onto the quaysides. On 11 July 1991, 08646 is seen returning from the quayside at Swansea Docks with a train of empty coal containers, which had been used to carry coal being exported to Ireland. I love the nonchalant poses of the crew, especially the shunter as he rides shotgun on the locomotive's steps. **Photographic Details: Canon AT-1, Ilford XP1, Canon 100mm f2.8 lens, 1/250 at f8.**

Below: As I continued to undertake research as a surgical registrar, I often travelled to attend scientific meetings or to give presentations and I made it a rule to always have a camera of some sort with me. Such was the case on 22 October 1991, when I had to present at a meeting in London. 47717 is seen at the buffer stops at Paddington, where I quickly snapped it before rushing off to the Royal College of Surgeons for an evening meeting. This image is also one of the first I took using the new Ilford XP2 film, which was introduced in 1991. **Photographic Details: Canon EOS100, Ilford XP2, Canon EF35–70 f3.5–4.5 lens, 0.5 seconds at f5.6, resting on the buffer stop.**

Above: Working at Frenchay Hospital from September 1991 to September 1992, I was often on the northern side of Bristol. Freight was far less prevalent in Bristol than it had been in the North East, so my after work photographic outings were far less frequent. I did however explore both the oil and waste terminals at Westerleigh, as well as Avonmouth Docks and the Tytherington branch. On 20 November 1991, 56031 "Merehead" stands at Westerleigh Waste Terminal, as waste containers from Avon County Council are loaded for onward carriage to Calvert. This service had started in 1985 and ceased in 2011. **Photographic Details: Canon EOS100, Ilford XP2, Canon EF35–70 f3.5–4.5, 1 second at f5.6, resting on a box.**

Right: I was delighted with this image, which is rare in so many ways. It was taken from the hard shoulder of the M1, heading north, on one of my many journeys between Newcastle and my parents' house in Cardiff. I had spotted the semaphore bracket several times before and on this day, 13 September 1991, I noticed workmen as we sped past, so pulled onto the hard shoulder, raised my bonnet and walked back to see what was going on. The M1 was significantly quieter in those days. As I got to the overbridge, 56020 crept into view with a Stanton Ironworks to Toton trip freight. I am not sure how much longer this traffic ran, but the permanent way staff were arranging to demolish the semaphore signals, because the next time we passed they had gone. During the Second World War, Stanton Ironworks was a major producer of armaments and iron production continued until 1974. The foundry was bought out by a French firm in 1983 and continued to produce iron products until 2007, when the last pipe was cast. The site is still rail connected and there are plans to reinstate it as part of a new industrial development on the site of the former ironworks. **Photographic Details: Mamiya 1000S, Fuji Provia, Sekkor 80mm f2.8 lens 1/250 at f.5.6–8.**

Left: One bonus of writing regularly for Rail Enthusiast was the opportunity to arrange visits to freight locations around the country. Having covered the demise of Speedlink in some detail, I was invited to be a guest of Railfreight Distribution (RfD) on the Dover to Calais train ferry. On 14 March 1991, 33205 and 33211 carefully reverse ferry vans onto the ferry in tandem, to avoid the ship listing to one side. At the time, RfD predicted the ferry would have a long life carrying cargos that would be banned from the forthcoming Channel Tunnel. In the event, the service ceased in 1994 when the Channel Tunnel opened and the tariffs for paths through the Channel Tunnel have all but prevented any rail freight from using the route. Despite there now being a railway line beneath it, there is now less rail freight across the Channel than there was 30 years ago! **Photographic Details:- Mamiya 1000s, Ektachrome 200, Sekkor 80mm f2.8 lens, 1/250 at f8.**

Below: As my time in Newcastle was drawing to a close, I decided to make one last day trip up to Scotland, as I reckoned it was unlikely that I'd get there again much, after moving to Bristol. I was fortunate to capture some of the last coal trains on the Bilston Glen branch to the south of Edinburgh, as stockpiles of coal were being cleared from the partially demolished colliery. I was also fortunate that the southern Edinburgh bypass had just been partially opened, creating a new photo location at Loanhead. On 2 May 1991, 26006 curves past the grain elevators at Loanhead with the 6B09 loaded MGR working from Bilston Glen to Millerhill Yard, where the train would be staged before onward haulage to either Cockenzie or Longannet Power Station. **Photographic Details: Mamiya 1000s, Fuji Provia, Sekkor 150mm f3.5 lens, 1/250 at f5.6–8.**

1992

Another year and more change! I was appointed as Lecturer in Surgery at the University of Wales Medical School in Cardiff, which was a promotion to Senior Registrar grade and much harder work. Unable to sell our house in Bristol, I commuted daily across the old Severn Bridge, leaving home every day at 06.00 and often not returning until 19.00 or later. On top of that, I found myself on a 1 in 3 on-call rota (83 hours a week) so not surprisingly, virtually my only British photography was from a couple of trips to take images for book and magazine projects. 1992 was also a momentous year in that I made my first trip to China with Dorridge Travel. I had never thought I would be able to afford a trip to China, as with a young family there was often barely enough cash left over at the end of the month for a meal out. For the previous five years, I had been saving all the income I received from magazine articles in a separate account and had amassed just enough for the trip to Zhongwei and Tonghua. In 1989, my mother, who also loved steam engines, had bought me a Colin Garrett book about China, which had put China "on my radar". Then, inspired by a fantastic picture taken by Peter Skelton which was published in Steam Railway early in 1992, I broached the subject with my then wife who agreed I could spend my "ill-gotten" authorship gains on the trip. Little did I know that I would be fortunate enough to make over 30 further trips to photograph Chinese steam before its final demise in 2021/22.

Above: One set of photographs that I was fortunate to take in 1992 was those of the briefly reinstated coal service to Padiham Power Station. Coal delivery had ceased in the early 1980s, but at the end of 1991, Trainload Coal instigated a service from Maryport Coal Disposal Point to the power station, effectively re-opening the short branch from Rose Grove. On 23 March 1992, on my way to photographing the last iron ore trains from Hunterston to Ravenscraig, I stopped off at Padiham and captured 60071 leaving the power station, on its way back to Rose Grove, where it would run round for the journey to Cumbria. This service ceased later that year.
Photographic Details: Canon EOS10, Ilford XP2, Canon EF100mm f2 lens, 1/250 at f5.6–8.

Above: Two days later, on the way back from Ravenscraig, I took the east coast route, stopping to chase the Redcar to Redmire limestone train. 60038 passes the derelict Leyburn station with the empty hoppers. The through line across the Pennines to Hawes had closed in 1964 and general goods traffic ceased in 1982. These limestone workings were nearing their end when this view was captured, the last of them running in December 1992. Leyburn station has since been transformed and is now the headquarters of the heritage Wensleydale Railway. **Photographic Details: Canon EOS10, Ilford XP2, Canon EF100mm f2 lens, 1/500 at f8.**

Right: On 24 March 1992, 60030 winds out from under the iron ore loader at Hunterston with the morning train to Ravenscraig. The ore terminal was able to upload 2800 tons of ore an hour from ships in the port. After Ravenscraig closed later in 1992, the terminal was used to handle imported coal for both Drax and Longannet power stations. This traffic ceased in 2016 and the tower and conveyors were demolished in 2019. Unlike so many industrial sites which end up as housing estates or leisure parks, the land is to be used to accommodate new wind turbines. **Photographic Details: Canon EOS100, Kodachrome 64 professional, Canon EF100mm f2 lens, 1/250 at f5.6.**

Above: Early in 1992, I received a phone call from long-time friend and railway film maker, Steve LeCheminant. He had been making a Telerail programme about Germany and filmed the narrow gauge line between Oschatz and Mugeln in Saxony. He waxed eloquent about the amazing little line that hauled standard gauge freight wagons to the china clay works at Kemmlitz, on top of "Rollerbruecken", which are narrow gauge wagons that accommodate the larger wagons on top. Often when told of an "unmissable" opportunity to photograph something, it plays on my mind until a plan is finally hatched to make the trip. So it was with Oschatz. I took three days leave and headed to Berlin where I picked up a hire car. Staying in the only hotel in Oschatz, which was surrounded by cobbled streets and had virtually no cars on the roads, my two-day visit was fantastic. In the event, freight traffic actually continued until 2001, so I need not have rushed. On 17 May 1992, Saxon Meyer 0-4-4-0T 099713 shunts in Mugeln Yard. The standard gauge wagons can clearly be seen to its left, sitting on their narrow gauge trolleys. **Photographic Details: Canon EOS10, Ilford XP2, Canon EF 200mm f2.8 lens, 1/500 at f11.**

Above: One of my few days out in 1992 took me to the Scunthorpe area. Hearing that it was being operated by pairs of 20s, I was keen to visit the Flixborough branch again. On 29 July 1992, 20104 & 20214 are seen passing Crosby Mines signal box with a trip working from Scunthorpe Trent Yard to Dragonby Sidings. It was sad to see that the box had been the victim of an arson attack just weeks previously, following its closure on 13 November 1991. **Photographic Details: Canon EOS100, Kodachrome 64, Canon EF 50mm f1.4 lens, 1/250 at f5.6–8.**

Left: I would never have managed to take this picture but for a chance encounter by the roadside. A young man was walking along the road, camera round his neck and clearly taking pictures of the steam line. I stopped to chat and he explained that one really needed a car to do justice to the line, but he didn't have one. He did however know every spot along the line, having walked it several times before Germany's reunification. I suggested that I drove and he could guide us to the good locations. We spent an enjoyable day together, which provided both of us with plenty of excellent pictures, including this one in the meadows outside Oschatz. Saxon Meyer 0-4-4-0Ts 099703 & 099705 make a stirring start from Oschatz as they head to Kemmlitz. During the day he gave one of the most telling observations about the recent reunification, which I have never forgotten. "The west has taken our steam engines and given us good cars, good cameras and sex shops. I'd rather have kept the steam engines!" **Photographic Details: Canon EOS100, Kodachrome 64, Canon EF 100mm f2 lens, 1/250 at f5.6.**

Above: The big railway event of 1992 for me was my first visit to China and this shot is as near as I ever got to recreating the magnificent image by Peter Skelton which inspired the trip. During the 1990s, China still had four classes of steam locomotive operating on its rail network, which were classified using a two letter code. The QJ Class engine shown here is a 2-10-2, 4715 examples of which were built between the 1950s and late 1980s. The other three classes, JS, JF and SY, were all 2-8-2s. Unfortunately, November 1992 was unseasonably warm so we didn't get the fantastic steam effects that I had hoped for. On 4 November 1992, China Railways QJ Class engines 6229 and 2647 round the famous curve at kilometre 721, west of Mengjiawan, late in the morning. It may have been unseasonably warm, but at dawn the temperature was well below freezing leading to one of the funniest experiences in my entire 50 years of photography. John Bennett, from El Vendrell, near Barcelona, had come on the trip having carefully worked out that this location was on the same latitude as Barcelona, so wore just a shirt and sleeveless cardigan. His immortal words when alighting from our tour bus this morning were "Bloody hell Rhodes, it's a bit parky". He hadn't read the tour notes and as an emergency measure myself and two other photographers buried him lying down in the sand, with just his head above the ground, placing a cardigan over his head to keep him warm! Later that day we purchased a bright pink puffer jacket for him, which would later save his life when we travelled to Tonghua, where the mercury dipped to minus 17 degrees. **Photographic Details: Canon EOS10, Kodachrome 64, Canon EF 50mm f1.4 lens, 1/250 at f5.6–8.**

Right: A second favourite from my first trip to China was this snowy shot at the Hunjiang Ironstone Railway. As mentioned above, the temperatures dipped once we headed to the North East of China, as did the frustrations. There were cancelled flights and multiple instances of getting lost, as steam tourism was relatively new in China in 1992 and nobody seemed to know where we wanted to go. As I recall, after driving round until nearly dusk, we finally found the operational centre for this charming little narrow gauge system just 30 minutes before it went dark, leaving little time for photography. As for John Bennett, he had missed the train from Changchun and had to take a 13-hour taxi drive through the snow to re-join the group! **Photographic Details: Canon EOS10, Ilford XP2, Canon EF 28mm f2.8 lens, 1/60 at f11.**

1993

1993 was a year dedicated mainly to surgical training in Cardiff and as such, I made very few outings in the UK to take photographs, apart from some opportunistic short visits to lineside locations in South Wales. One big trip that year however, was to California. I had been awarded a Pancreatic Society travelling fellowship to visit the University of California in San Diego and while there, I managed a weekend exploring Barstow, Cajon Pass and Tehachapi; locations that I have returned to several times since.

Left: On 15 February 1993, I was allowed an official visit to Margam steelworks, in order to prepare for an article I was writing for Rail magazine on the site's internal railway network. It was a real privilege to be allowed access to all areas, including here inside the Hot Rolled Coil Bays. Brush Bagnall Bo-Bo diesel-electric 906 is seen shunting BAA and BBA bogie wagons which are being loaded with coils bound for secondary steel plants all over the UK. **Photographic Details: Canon EOS10, Ilford XP2, Canon EF50mm f1.4 lens, 1/30 at F4.**

Below: Not having a detailed map of the area, my pictures of the Tehachapi Pass in California were not the best. This view was taken from the hard shoulder on the Interstate 58 at Cable. Eight locomotives from Classes SD45, SD40, GP50, GP60 and B40, numbered 5835, 5004, 4029, 7441, 3811, 3813 3841 & 4032 head train 991, a hotshot (express freight) intermodal from Richmond to New York. There were two additional locomotives in the middle of the consist (Nos. 5123 & 5054). Later visits led to much better images along the Woodford Road, but my first visit certainly whetted my appetite for American Railroads. **Photographic Details: Canon EOS10, Kodachrome 64, Canon EF50mm f1.4 lens, 1/500 at f4.**

Right: Whilst on summer holiday in a cottage in the Yorkshire Dales, I managed just one decent railway photograph all week! On 10 August 1993, 60016 approaches the south end of the Rylestone branch, where it passes under the Skipton bypass. It is hauling stone destined for the terminal at Dairycoates in Hull. **Photographic Details: Canon EOS10, Kodachrome 64, Canon EF100mm f2, 1/250 at f5.6–8.**

Below: On 11 November 1993, 08790 heads to the quayside in Swansea Docks with rebranded Cawoods containers which are carrying export coal for Ireland. The bridge over the railway between Swansea Burrows Sidings and the dock complex on the A483 was a favourite location of mine for many years. By the time of this view, most of the cranes around the dock had been removed, the tracks in the docks rationalised and the 100-lever King's Dock Junction signal box, which had stood above the fourth and fifth wagons of this train, had been closed and demolished in 1987. On the other side of the bridge, the 110-lever Burrows Sidings signal box had closed at the end of 1990. A return visit to the same spot a couple of years ago was shocking; trees obscured the view of what remains of Burrow Sidings to the east and the docks to the west were well on the way to becoming a pleasure marina and residential area, with no obvious sign of it being a commercial port. **Photographic Details: Canon EOS10, Kodachrome 64, Canon EF100mm f2, 1/250 at f5.6–8.**

1994

On 31 December 1993, my family boarded BA21 and headed for Brisbane, where I spent a year as the Lecturer in Surgery at The Royal Brisbane Hospital. Landing in Brisbane around midnight on 1 January, in the middle of a heatwave, where the temperature was 35 celsius, we wondered what had hit us. The year was the most intense in my surgical training and I learned more than I could have imagined. Most of my railway photography took place in Australia that year and presenting papers at meetings in Hobart, Adelaide and Perth allowed me to see a fair bit of the country. We even got treated to a week in Sydney at the end of the year by one of my bosses, George Fielding. I had to make three long-haul trips during the year; I visited Japan and the USA to present research work and had to return to the UK in August of 1994 to look at a couple of consultant jobs – my cameras travelled everywhere with me.

Above: Just to the west of the Royal Brisbane Hospital were the Normanby Park Yards. These were a five-minute walk away, so at lunchtime I could pop over with a sandwich and see what freight traffic was about. In February 1994, a couple of EMD Co-Co locomotives shunt containers. Queensland Rail had always purchased English Electric locomotives and the older 1200 Series Co-Cos were based on the English Electric Class 20. These two locomotives were very similar externally to the older 1200 Class but had EMD engines. The yards here have now been subsumed by development. **Photographic Details: Canon EOS10, Kodachrome 64, Canon EF100mm f2 lens, 1/250 at f8.**

Top Right: A family holiday to Heron Island necessitated taking the ferry from Gladstone to the island on the Barrier Reef. Gladstone, unbeknown to me, was and remains one of the largest coal exporting ports in Australia, along with Hay Point to the north and Newcastle in New South Wales. Two empty coal trains stand in the extensive yard at Gladstone waiting to return to the coalfields. On the right, 2100 Class 2157 is bound for the mines near Baralaba on a non-electrified branch and to its left 3500 Class 3544 heads for the mines around Emerald. **Photographic Details: Canon EOS10, Kodachrome 64, Canon EF200mm f2.8 lens, 1/250 at f8.**

Bottom Right: On 28 August 1994, I had arrived back at my parent's house in Cardiff, for a flying visit to look into two consultant posts. A quick trip to Cwmbargoed found 37701 & 37898 waiting to depart for Aberthaw Power Station with their load of coal. At the time of writing in 2021, Cwmbargoed remains one of only two places where domestic coal is still moved by rail in South Wales, as it supplies coal to Port Talbot Steelworks. It has always puzzled me how this is seen as less green than importing coal from thousands of miles away when the journey to Port Talbot is mainly downhill and just under 50 miles from here? **Photographic Details: Canon EOS10, Fuji R100, Canon EF100mm f2 lens, 1/250 at f8.**

3544
3544
2157
2157

701

Above: The Australian Railway Historical Society ran several weekend specials during 1994, including this one on 19 September 1994 from Brisbane to Toowoomba. "Baby Blue", a DD17 Class 4-6-4 tank engine, was used on Brisbane's suburban services during the 1950s and 1960s, before being preserved by Queensland Railways' heritage department. Here it leaves Murphys Creek on the climb up the ranges to Toowoomba. The main line from Brisbane to Toowoomba climbs a height of 1211 feet in just 16 miles and this was the first narrow gauge main line in the world when it was built between 1865 and 1867. Due to the route's many twists and turns, it is exceptionally difficult to photograph and I failed to find any decent locations during my year in Australia. **Photographic Details: Canon EOS10, Kodachrome 64, Canon EF 28mm f2.8 lens, 1/250 at f8.**

Bottom left: In order to take up the gift of a week in Sydney, as mentioned above, we drove from Brisbane, and stopped off in Newcastle. The steelworks there were still busy in 1994 (steel making ceased in 1999), as were the coal export terminals at Port Waratah and Koorangang Island. 9000 Class 9005 was brand new when this view was taken on 22 November 1994, being one of 35 locomotives leased that year by Clyde Engineering to Freight Rail in New South Wales, on a "power by the hour" basis. The locomotive is pulling through the unloading bunker at Port Waratah, with coal from the Hunter Valley that is bound for China. **Photographic Details: Canon EOS10, Kodachrome 64, Canon EF100mm f2 lens, 1/250 at f5.6–8.**

Below: No selection of images from 1994 would be complete without one from my first visit to Japan, to attend the 4th World Congress of Endoscopic Surgery in Kyoto. The railway station in Kyoto was opposite my hotel for the week and proved invaluable as a source of food. I had flown from Brisbane, via Cairns and had changed 100 Australian dollars into Yen, assuming this would be enough for food for the week. I arrived at my hotel and ordered a burger and fries only to find that it had cost me the best part of 50 dollars! My bosses were kind in giving free tickets to dinners and taking me out, but salvation came every night when dozens of tents appeared on the forecourt outside Kyoto station, selling soup and chicken skewers. A full meal cost just two dollars at the stalls, which were packed with workers heading home. By the next morning the stalls had all been cleared away for another working day. A bullet train between Tokyo and Osaka is pictured entering Kyoto station on 16 June 1994. **Photographic Details: Canon EOS10, Kodachrome 64, Canon EF200mm f2.8, 1/250 at f5.6.**

1995

Returning to the UK at the end of 1994, the first half of the year was spent finishing my surgical training in Cardiff, before taking up a consultant general surgery post in Norwich. This signalled a fundamental change in my photographic activity because almost every waking hour in the UK was taken up with my new job, leading to a drastic fall in my UK photographic output. I have called this my "Morrison Gap" which is in honour of Brian Morrison, who had a similar 15-year period in his life, when he took very few photographs due to work and family commitments. The difference for me was that my work often took me overseas to lecture and I also started travelling to China regularly to photograph steam. I did continue to take pictures in the UK, but between 1995 and 2010 they make up only a small proportion of my railway photographs. Those who dislike steam and overseas images should perhaps skip to the chapter covering 2011!

Above: On 27 April 1995, I was invited as resident doctor and a German speaker on the first Steam Railway magazine tour to Wolstyn in Poland. Steam Railway chartered a GB Airways Boeing 737 during its down time in the middle of the night and we flew to the military airport at Poznan. Arriving at 02.00, I recall two bewildered soldiers who basically counted us off the plane and presumably counted us back on a couple of days later. Having assumed there would be food shops and cafes in Poland, I spent most of the two-day trip starving, as even in our hotel in Poznan we were rationed to one slice of bread each! How things have changed – during my 2007 visit to Wolstyn we stayed in the town and ate steak every night at an Argentinian restaurant which served excellent wine too! Having fallen asleep in the locomotive depot at 04.00, I roused myself to capture Class Ty45 2-10-0 45-379 leaving Wolstyn with the 06.35 departure to Grodzisk, which consisted of both passenger and freight vehicles. **Photographic Details: Canon EOS10, Ilford XP2, Canon EF200mm f2.8 lens, 1/250 at f8.**

Above: This image is unique in the book in that it is a print scan. Having made the acquaintance of Gene Nocon during my 1993 pancreatic fellowship in San Diego, I sent the occasional selection of negatives for him to print for me when I had failed to get decent results myself. He had worked with David Bailey and Norman Parkinson in the swinging sixties and acted as a photographic tutor to Prince Andrew and Koo Stark. His unique darkroom methods yielded superb results and I revisited him several times, hoping some of his magic would rub off. He sadly passed away in 2011. The picture is a sunrise shot at Avon Yard in Indianapolis taken on 12 May 1995. **Photographic Details: Canon EOS10, Ilford XP2, Canon EF50mm f1.4 lens, 1/500 at f11.**

Below right: Between finishing my training and taking up my consultant post, I had a few weeks off and visited East Germany for a couple of days. The main aim was to document as much of the former DDR (German Demoncratic Republic) infrastructure as possible, as well as to produce to an article for Rail magazine. This image of "Ludmilla" Class 232 diesel-electric 232269 was taken on 11 July 1995 and typifies scenes at the time. A block train of lignite is seen arriving in Gera, where it will reverse into the power station in the background. Even today, 40% of Germany's electricity is generated from coal (compared to 0% in the UK) and most of the coal used is "Braunkohl" or lignite, which has low calorific value and is very polluting. To the right is a hump that leads to the goods yard at Gera, which was equipped with two electromechanical retarders. There was also a sizeable engine shed with a turntable. Fast forward a decade and all of this had been swept away, with no freight yard or depot and just the occasional railbus calling at the impressive Gera station. **Photographic Details: Canon EOS10, Fuji R100, Canon EF100mm f2 lens, 1/250 at f8.**

Above: The next four images from 1995 reflect the outings I made in South Wales, to take as many pictures as possible, before I left for Norfolk. On 29 March 1995, having been appointed to Norwich, earlier that month, I revisited a favourite spot on the new Cardiff Docks link road. 60096 is seen passing Pengam Freightliner terminal with a Margam to Ebbw Vale steel coil train, whilst 47114 waits in the goods loop, with a Speedlink service from Cardiff Tidal Sidings to Willesden Yard. The Freightliner depot in the background has since been redeveloped as housing, despite it being on the River Rumney flood plain and container services now use a new depot at Wentloog, halfway to Newport. **Photographic Details: Canon EOS10, Kodachrome 64, Canon EF100mm f2 lens, 1/250 at f5.6.**

Right: A favourite spot over the years from 1982, when I first spent an afternoon in the signal box there, was Park Junction. The signalman often let me roam around the area, as I did on 6 July 1995. 60034 winds over the junction with a Margam to Ebbw Vale steel coil train. It was in 1995 that tin plate production ceased at Ebbw Vale and then in 2001 the plant closed completely. The massive plant had stopped producing steel in 1978 and the 1995 tin plate works closure led to the loss of the last 780 steel-related jobs in Ebbw Vale. Passenger services have since returned to the valley and the site of the former steelworks has been redeveloped with housing, a hospital and a wetland park. **Photographic Details: Canon EOS10, Fuji Provia, Canon EF200mm f2.8 lens, 1/250 at f5.6–8.**

Above: During my last week in South Wales, I spent a glorious summer evening in Barry. On 24 July 1995, the former Burry Port & Gwendraeth Valley Railway shunter 08995 "KIDWELLY" leaves the dock branch with tanks which will form the Mondays, Wednesdays & Fridays only Barry to Baglan Bay oil train. My notes record that the 08 had made a trip down the branch at 16.00, only to return light engine because the tanks were not ready. They were taken forward from Barry Yard by 37906. **Photographic Details: Canon EOS10, Fuji Provia, Canon EF 50mm f1.4 lens, 1/250 at f8.**

Below: A second picture taken on 24 July 1995 shows 56115 curving off the Barry Docks line towards Cadoxton, with the Mondays, Wednesdays & Fridays only chemical working from Dow Chemicals in Barry Docks to Burn Naze on the former Fleetwood branch. The road that begins above the four cars on the right and which crosses the railway where the rear tanks are seen, was the route to Barry Yard taken by westbound traffic from the docks. Barry Docks opened in 1888 and by 1913 was the largest coal exporting port in the world, handling over 11 million tons of coal each year. Its decline has been steady but relentless. In the 2000s, a large housing and leisure redevelopment was built on much of the area, then a solar farm generating 4.5 MWh of electricity was added. Cargo handled at the port has now dwindled to less than 200,000 tons per annum. Like so many docklands, the area is rapidly being turned over to leisure, rather than trade and industry. **Photographic Details: Canon EOS10, Canon EF 50mm f1.4 lens, 1/250 at f5.6–8.**

1996

Looking at my logbook, I had just two days out taking railway photographs in the UK during 1996. I did however discover the delightful Brief Encounter restaurant at Wymondham station, which was to become a firm favourite until the owner was forced to sell it in 2011 and it had to relinquish the name due to a lawsuit from the copyright holder of the name Brief Encounter. To eat a curry, sitting on old first class carriage seats, with nameplates and timetables on the wall and a lovely open fire was marvellous. I made my second visit to China in 1996, beginning a tradition of at least one trip each winter for the next 15 years.

Below: I've always called this "the ice skating shot". There is more back story to this image than meets the eye. It was our first joint trip to China (Rhodes, Murphy & LeCheminant) and we were once again guided by the late Hugh Ballantyne who was working for Dorridge Travel. The first main location we visited was Chengde, where the steelworks branch climbed spectacularly from the China Rail interchange to the steelworks. After this location, the railway ascended and trains needed to be double-headed and banked by another locomotive. This picture of SY Class 0872 was taken as we arrived in Chengde on 20 January 1996. We spent the following day north of the banking station in Chengde, but our guide, Hugh, mysteriously disappeared with the tour bus mid-morning for important business. Little did we know that he had decided to return to the frozen river when the light was better and there were more skaters, to take some more views on his own. We only discovered this when his image from 21 January won the Amateur Photographer competition, the first prize of which was a Mamiya camera! My image is nowhere near as good as Hugh's prize-winning shot, but I feel a bit grumpy to this day that we didn't get to join him back at the river. **Photographic Details: Canon EOS10, Fuji Provia, Canon EF50mm f1.4 lens, 1/250 at f2.8.**

Right: One of many great shots we achieved at Yebaishou during our January 1996 China trip was this picture of QJ Class 6418 climbing from Yebaishou towards Chifeng, with an empty coal train returning to one of the many mines in the Chifeng area. I especially like the indifference of the cyclist and his child as they trundle down the road. **Photographic Details: Canon EOS10, Kodachrome 64, Canon EF100mm f2 lens, 1/250 at f5.6.**

Right: Having taken a few images on the hillside above Chengde, where Vernon and I discovered that we had both been to the same Cambridge college, Gonville and Caius, and had both won the Cambridge University Railway Club photographic contest three years running, we decided to wander down the line in search of new locations. We arrived here at the "banking station". We were very fortunate to capture this image, as the building behind the train was apparently a prison and previous photographers had been arrested at this location. China Railways JS Class 6217 starts to move its train, with a pair of SY Class banking locomotives seen on the rear, whilst the yard pilot, JS Class 6018 simmers to the right. This photograph was taken on Kodachrome 64 film, but the resulting colour image was almost monochrome, so after scanning I have converted it to black and white using Photoshop. **Photographic Details: Canon EOS10, Kodachrome 64, Canon EF200mm f2.8 lens, 1/250 at f5.6**.

Above: Anshan steelworks was a favourite with Western rail enthusiasts, but in our 1992 visit, a bizarre game of Chinese cat and mouse took place, whereby we spent two hours circling the perimeter of the works, only to be allowed into the locomotive depot as the light disappeared! By 1996, visits were more frequent and access was granted to particular locations, including the slag tipping site pictured here. SY Class 0130 draws its ladle wagons forward in temperatures that were so cold that the water leaking from the pipes on the site had frozen, giving this dramatic view of fire and Ice. The ladle wagons illustrated had a habit of slopping molten metal if they were rough shunted and a few months after this image was captured, an over-enthusiastic Japanese photographer got too close to one of these trains and ended up with molten metal on his head, causing his inevitable demise. Needless to say, visits to the steelworks were prohibited thereafter. **Photographic Details: Canon EOS10, Kodachrome 64, Canon EF50mm f1.4 lens, 1/250 at f8.**

Above: My second UK day out that year was on 23 July 1996, when I visited Immingham and Barnetby. Unusually on this day, trains were being routed out of Immingham Docks using the former Immingham Light Railway route via Grimsby, due to a line blockage to the west. This provided the rare sight of 60038 passing Pyewipe Road level crossing with the 6D85 freight from the Tioxide plant to Roxby Landfill Site on the Dragonby branch. I remember the signalman coming out of the box to accept the single line token looking very dishevelled with his braces down, then seeing me with my camera, he suddenly put his braces back over his shoulders! **Photographic Details: Canon EOS10, Fujichrome R100, Canon EF50mm f1.4 lens, 1/250 at f4.**

Bottom left: One of my only two days out from Norwich to take photographs that year was an excellent day visiting the Calder Valley on 3 April 1996 with fellow photographer David Pearce. This image was fortunate enough to make it onto the cover of Rail magazine, something I have very rarely achieved during my 50 years of photography. 60008 winds out of the now-closed Eggborough Power Station, returning to Askern Colliery to pick up more coal. **Photographic Details: Canon EOS10, Fujichrome R100, Canon EF50mm f1.4 lens 1/500 at f5.6.**

Below: On 29 July 1996, 60016 is captured passing through Wymondham station with the daily Mountsorrel to Norwich stone train. With the arrival of the excellent Cavalex 00 gauge Redland stone hoppers and the Hornby Class 60, this train can be perfectly replicated on Barrow Hill Junction model railway, bringing back many memories of toasted tea cakes and coffee at Brief Encounter on Platform 1 at Wymondham station. **Photographic Details: Canon EOS10, Kodachrome 64, Canon EF50mm f1.4 lens, 1/250 at f5.6.**

1997

Work dominated 1997; apart from a family holiday in the Lake District and a couple of days out, very few UK photographs were taken. Visits overseas were predominantly short trips to lecture, teach or attend courses and I even made a couple of "day trips" to the USA! The small selection of images for 1997, reflects this busy year when photography took a back seat. There was a visit to China and my favourite photograph from that trip has been included.

Above: I made a day trip on 9 April 1997, visiting Drakelow Power Station, Calverton Colliery, Bestwood Park and Thoresby Colliery. Sadly, the semaphore signals at Drakelow had been demolished just a couple of weeks prior to my visit and these lay mangled on the ground next to the old signal box. On the bright side, I managed to capture a train arriving at the power station and then leaving. After having brought coal from Denby Colliery, 58022 leaves the yard at Drakelow with an empty MGR working to Toton Yard.
Photographic Details: Canon EOS1n, Kodachrome 64, Canon EF50mm f1.4 lens, 1/250 at f5.6.

Below: During a holiday in the Lake District, a hike along the River Lune at Tebay seemed like a good idea! This classic location favoured by the photographer Peter Robinson was duly identified and 90135 was photographed on the 6K85 wagonload service from Carlisle to Bescot. It was pleasing to capture what was effectively a Speedlink service six years after the closure of the wagonload network. **Photographic Details: Canon EOS1n, Kodachrome 64, Canon EF50mm f1.4 lens, 1/500 at f4.**

Above: A real coup during the summer holiday was the unexpected chance to photograph the short-lived containerised milk train, which was an experiment by the freight operator DRS. The Milkliner, as it was branded, travelled between Penrith and Cricklewood. It is seen here behind 20303 & 20304, passing Hest Bank, photographed whilst the children played on the beach! A well-known railway photographer was in charge of trying to establish this service at the time and a detailed account of his struggles makes sad reading, as the service was abandoned just weeks after this photograph was taken. **Photographic Details: Canon EOS1n, Kodachrome 64, Canon EF28mm f2.8 lens, 1/500 at f4–5.6.**

Following page: Our January 1997 trip to China was not as successful as it might have been, but we became some of the first photographers to visit the legendary Jinpeng Pass. As one of the earliest groups to visit the area, we took some time to get to grips with the many twists and turns along the line, as it looped back and forth over the pass. On our last day there, 21 January 1997, we finally got everything right for a shot on the second level above Reshui and this picture of a pair of QJ Class 2-10-2s on a westbound freight became the cover shot for our first China book, Extreme Steam. There was much jumping around and fist pumping after the train had passed, as we had finally nailed a decent Jinpeng shot! **Photographic Details: Canon EOS1n, Fuji Provia, Canon EF200mm f2.8 lens, 1/250 at f5.6–8.**

1998

The years began to develop the same pattern – work, work, work and then one or two trips to China. Most UK photography was undertaken close to home in Norwich or on the occasional day out. In addition, about once a year I produced a long article for Rail magazine, which prompted me to make a trip to research and illustrate the piece.

Above: Through contacts at the Mid Norfolk Railway, I was fortunate enough to receive advance notice of the military specials, which ran on the main line network to Wymondham and then continued along the heritage Mid Norfolk Railway to Dereham. I'm not sure how many of these trains ran, but I seem to have photographed half a dozen of them bringing equipment to the military base at Swanton Morley. On 26 September 1998, 47363 is seen on the tail end of the train at Dereham, having brought the special from Didcot (I think) to Wymondham. At Wymondham, it reversed and was taken forward by one of the Mid Norfolk Railway's diesel locomotives. A long lens was needed to keep away from the soldiers guarding the train as it was unloaded! **Photographic Details: Canon EOS 1n, Ilford XP2, Canon 300mm f4L lens, 1/500 at f11.**

Right: Radyr was always a favourite location of mine, ever since I first cycled there in 1976. In 1998 major changes were afoot, as part of the new Cardiff Circle Line service from Coryton to Radyr via Danescourt. On 16 April 1998, 60063 pauses at Radyr station with a Tower Colliery to Aberthaw Power Station MGR working. To the left of the train, the station's new third platform is under construction and the splendid semaphore signals will only remain in use for a few more weeks. The 107-lever Radyr Junction signal box which controlled these was closed when the area was resignalled on 12 May 1998. **Photographic Details: Canon EOS1n, Kodachrome 64, Canon EF100mm f2 lens, 1/125 at f5.6.**

On 5 December 1998 we "discovered" the colliery yard and railway at Hengshan, near Jixi in China. The process of our arrival there was tortuous and began with a conversation with our guide one year earlier. In January of 1998, we had visited the loop at Modaoshi and were informed that there were a lot of QJ Class locomotives allocated to a big depot at Jixi. It was because of this that eleven months later we had retuned to Jixi. The local guide in Jixi then suggested to our guide Li Weishu that Hengshan was worth a visit because there was "a colliery railway" there. I think we were the first western enthusiasts to make it to Hengshan, which of course went on to be a frequently visited location for the next decade. At sunset on our first day there, China Rail QJ Class 6147 indulges in a bit of shunting, whilst a colliery SY Class simmers in the distance. Twice a day, China Rail "top and tailed" heavy coal trains from Hengshan to the main yard at Jixi Xi. **Photographic Details: Canon EOS 1n, Kodachrome 64, Canon EF200mm lens, 1/250 at f11.**

1999

The year was marked by some interesting local developments in the Norwich area. A Class 55 "Deltic" was utilised on express trains to London and local services to Yarmouth were handled by a Class 50 for a few weeks. My photography was restricted to local outings, with the exception of the usual visits to see my parents in Cardiff and one day out to the newly commissioned Wembley Yard. This was fascinating and the yard was exceptionally busy. Other than that, I made a trip to China, visited America to present research work with my team and made my first marshalling yard trip to the USA in preparation for the 2003 publication of the first edition of North American Railyards.

Above: With the signals gone at Radyr and the number of active collieries in South Wales down to single figures, I had to venture a bit further afield for my South Wales freight fix. Abercynon was a favourite spot, with its 93-lever signal box surviving until 2008. Here the signalman takes the single line token from the driver of 66048, as he pauses for a chat on his way from Tower Colliery to Aberthaw Power Station. **Photographic Details; Canon EOS1n, Ilford XP2, Canon EF100mm f2 lens, 1/250 at f8.**

Top right: On 15 July 1999, I found myself in Brownsville, Pennsylvania! The idea which took seed back in 1989 had germinated and I took a week of annual leave to visit and photograph as many hump yards in the east of the USA as I could. My notebook records that between 13 and 17 July 1999, I visited the yards at Oak Island in New York, Cedar Hill, East Deerfield, Selkirk, Syracuse DeWitt, Altoona, Cumberland, Conway, North Bessemer in Pittsburg, Willard in Ohio, Kirk in Indiana and finally Gateway & Maddison in East St. Louis. Somewhere along the way, I found myself seeing signs to Brownsville and vaguely remembered that there was street running there. Conrail GE Dash 8-40CW 8407, plus a second locomotive, rumble onto the main street with several thousand tons of coal. **Photographic Details: Canon EOS1n, Ilford XP2, Canon 70–200 f2.8L lens, 1/500 at f11.**

Below: The optimism for rail freight through the newly opened Channel tunnel was immense. Fortunate to be allowed an official visit by EWS, Paul Shannon and I made one of our last ever joint visits to Wembley Yard on 22 July 1999. We were allowed up the lighting gantries (at our own risk!) and after a busy afternoon shift, we spent a couple of hours in the yard after dark. 92023 awaits departure with 6B45, an evening Wembley to Dollands Moor freight. It was envisaged at the time that freight would be so heavy that there might be hourly shuttles to Dollands Moor – how mistaken these predications were. To the left is sister locomotive 92021.
Photographic Details: Canon EOS1n, Ilford XP2, Canon EF100mm f2 lens, 1 second at f8 using a tripod.

Previous pages: When this image was taken on 16 December 1999, we assumed that freights leaving Hunjiang to the north were bound for Baihe, but publication of a third China Rail Atlas by the Quail Map Company in 2008, revealed there was a mine just a few kilometres north at Zhazi, which is almost certainly where this long load of coal empties is headed. JS Class 5371 crosses the Hongtuya River (a tributary of the Hun River), which has an interesting melt pattern caused by the effects of warm water released by the local power station, after the river water has been used for cooling the generators. This view is now completely obscured by a four-lane expressway. **Photographic Details: Canon EOS1n, Fuji Provia, Canon EF50mm f1.4 lens, 1/500 at f4–5.6.**

Right: Our 1999 trip to China took in the Jinpeng Pass, the narrow gauge logging line at Zhanhe and finally a second visit to the Tonghua area. On 9 December 1999, QJ Class engines 6389 & 6351 catch the last rays of sun as they exit Tunnel 1 on the climb from Jinpeng to the summit. When there is a roaring gale in the valley, it really is a complete lottery as to where the steam will end up on a photograph. **Photographic Details: Canon EOS1n, Fuji Provia, Canon 70–200mm f2.8L lens, 1/250 at f5.6–8.**

Below right: The Zhanhe narrow gauge logging railway was one of the most northerly in China. During our visit in December 1999 the mercury dipped below minus 30 degrees. Here on 12 December we made a preliminary foray along the start of the 258 km system and managed to catch a single empty log train at sunset as it approached Erdaohe. The next day, after getting half way up the line in a small railcar, we had to turn back because of traffic congestion further up the line. This was a painful experience as the temperature inside the railbus was a steady minus 16 degrees all day – we effectively spent 12 hours crammed into a domestic freezer! It would not be until March 2001 that we eventually made it to the end of the line at Beiying by which time the system was on its last legs due to the timber in the forests having been over logged. The whole system closed in 2002. **Photographic Details: Canon EOS 1n, Fuji Provia, Canon 70–200 f2.8L lens at 200mm, 1/250 at f8.**

Below: During January of 1999, there was the remarkable sight of a Deltic working daily between Norwich and London Liverpool Street. Ostensibly this was due to a motive power shortage, but I have my suspicions that there may have been an enthusiast somewhere in the mix who made this happen. D9000 "ROYAL SCOTS GREY" stands at the buffer stops at Norwich on 13 January 1999, having just arrived with the 17.00 departure from Liverpool Street. **Photographic Details: Canon EOS1n, Kodachrome 64, Canon EF28mm f2.8 lens, 1 second at f4, resting on the buffers.**

The 2000s

2000

On a local front, 2000 saw the latter days of the Royal Mail trains from Norwich to London and some more military specials ran to Dereham. Otherwise, not much was captured on film in the UK. 2000 marked my first experience of a "Plandampf" which is a German innovation where regular services see steam locomotives replace the normal traction. I was persuaded to join Vernon Murphy for a Plandampf organised by the local Rhein-Pfalz government to celebrate 125 years of the Quiechtalbahn, with steam working based around Pirmasens. I managed two trips to China and while there, I "discovered" one of my favourite spots at Da'an Bei, just before the diesels arrived – more of that in the captions.

Below: We were so concerned that we would be locked out of our hotel in Pirmasens, that a late-night dash from Frankfurt airport was made in a hire car, when Vernon hit 180 kph (112 mph) and knocked off the wingmirror on the passenger side of the vehicle. The next morning, 1 October 2000, was misty when DRC Class 01 "Pacific" 01 118 broke through the morning mist as it left Pirmasens Nord Pbf with a train to Landau carrying over ten times the normal number of passengers due to the celebrations of the weekend. **Photographic Details: Canon EOS1n, Fuji Provia, Canon EF300mm f4L lens, 1/500 at f5.6.**

Above: Frankenstein Castle dominates this view of an express from Kaiserslautern to Neustadt and Mannheim on 3 October 2000. German Class 01 "Pacific" 01519 burst out of the tunnel to the sight of dozens of lineside photographers. It is amazing that so many photographers were allowed on the ballast and up the catenary supports, bringing step ladders and all manner of paraphernalia on this busy electrified main line. I could only get a picture by kneeling down on the ballast, to keep out of the way of other photographers who had arrived before Vernon and I and hope that a train didn't come hurtling from behind me, throwing up ballast, or worse, emptying the toilets! **Photographic Details: Canon EOS1n, Fuji Provia, Canon EF 100mm f2 lens 1/500 at f5.6.**

Insert above: During the Plandampf, Vernon and I were standing on an overbridge near Pirmasens when a black Mercedes pulled up and two Japanese photographers leapt out. They did a double take of Vernon and I and then bounced up to us demanding selfies with us – they said "Konnichiwa, Extreme Steam Team." It would seem our recently published book about China had made it all the way to Japan and had been well received there!

Left: The last shot of the day for us on 5 December 2000 captures the essence of Da'an Bei. All freight trains changed engines in the yard next to the station. In December 2000, the exchange was often from diesel to steam or vice versa. Just before rushing to the station to catch the train to Ranghulu, where we would change for Harbin and then catch a sleeper to Shungyashan for the first of our three visits there, we captured QJ 6825 with a mixed freight to Taipingchuan and Tongliao. This engine was later to work out its days on the Jitong line. **Photographic Details: Canon EOS1n, Fuji Provia, Canon 70–200 f2.8L lens, 1/250 at f8.**

Left: In March 2000, Vernon and I had discovered Da'an Bei, a busy junction four hours north of Changchun in China's Jilin Province. This came about after the mother of Li Weishu, our guide in China between 1996 and 2004, had been chatting to a retired engineman who told her (as a result of Li asking for information) that steam engines popped up in Changchun a couple of times a day from Da'an Bei and that there was a big steam shed there. This had an allocation of 47 QJ Class 2-10-2s and 9 smaller JS Class 2-8-2s. Later recounting details of this successful visit to Steve LeCheminant, who wasn't with us in March, led to a return trip in December 2000. Sadly, by then the steam fleet had been reduced to about half the size, but the weather however, was superb for photography. QJ Class 6887 is seen crossing the Songhua River with a freight from Ranghulu at sunrise on 5 December 2000, when the mercury reading was 24 degrees below freezing. The railway and a number of spectacular bridges cross the Songhua River in Harbin and Jiamusi, which is frozen for three to four months every winter. The barges park up for the cold months and their crews drill holes in the ice to fish and also to use as impromptu toilets! During our March 2000 visit here, a lorry fell through the melting ice whilst crossing the river, killing the driver. We had driven across this in our van just the day before! **Photographic Details: Mamiya 1000S, Fuji Provia, Sekkor 80mm f2.8 lens, 1/250 at f8.**

Left: The travelling post office (TPO) train from Norwich to London ceased operation in 2003 as I recall. The novelty of posting a letter directly into a letter box on the sorting carriage never wore off. With work exceptionally busy in 2000, the best I could do to record the UK scene was capture the last of the 47s at Norwich on the night mail to London, before they were replaced by Class 67s. 47767 awaits departure on 24 March 2000. **Photographic Details: Canon EOS1n, Fuji Provia, Canon 50mm f1.4 lens, 1 second at f4.**

Below: One of my favourite pictures on the Mid Norfolk Railway was taken on April Fools' Day 2000. 50019 heads up Haddington Bank with army personnel carriers bound for the military base at Swanton Morley, via the goods yard at Dereham. On the rear of the train was 66158, which had brought the consist from Didcot. **Photographic Details: Canon EOS1n, Fuji Provia, Canon 50mm f1.4 lens, 1/250 at f8.**

Above: My final image from 2000 is unusual in that I can take very little credit for it. We had arrived at Weihe in China's Heilongjiang Province on 8 December 2000 and I was feverish with a cough and beginning to hallucinate. It turned out that I had a very bad chest infection, which kept me confined to bed in Weihe, with a two-bar heater and food (including a tin of peaches as I recall) brought to me by Vernon and Steve. As we wandered round the market at Xiping, which is on the logging line from Weihe, Steve noticed that I wasn't taking pictures and was looking a bit funny – I told him I really didn't feel well at all, as I was having rigours and sweating profusely, even though it was minus 20 degrees. Steve put his hands on my shoulders and said we need some "cut aways" – point your camera here and take this shot with the shoes. So here it is, shutter pressed by me, direction and framing by Steve! This shows a China Railways C2 Class 0-8-0 working the afternoon Yulin to Weihi passenger service. **Photographic Details: Canon EOS1n, Ilford XP2, Canon 50mm f1.4 lens, 1/125 at f11.**

Top right: One or two local photographers had extolled the virtues of the locations to be found using a boat on the River Yare around Old Lakenham. I didn't have a boat, but I got wet feet trying to find a decent location from the river bank. 86235 crosses Lakenham Viaduct, which spans the Norwich to Ely line, with a London to Norwich express during July 2001. **Photographic Details: Canon EOS1n, Fuji Provia, Canon 50mm f1.4 lens, 1/500 at f5.6.**

Right: On 5 December 2001, when the dawn temperature was minus 14 degrees, we climbed up the hill above the level of the pagoda at Lindong in China. QJ 6884 leaves Lindong at 08.00 and heads for Daban with a mixed freight that includes coaches for the Jitong line which had been refurbished in Changchun. The white and blue coaches of the overnight Hohot to Tongliao service can be seen in the station platform on the far right. **Photographic Details: Mamiya 1000S, Fuji Provia, Sekkor 80mm f2.8 lens, 1/250 at f5.6.**

2001

I suspect the reader may become weary of the annual summaries for this period. I took virtually no images in the UK during 2001. The best I could find for the selection was the image I have chosen on the outskirts of Norwich. In contrast however, there were some professional successes with our research, winning some prizes in the USA and Europe and even getting featured in the New York Times. In addition, I undertook the now customary two trips to China with Vernon Murphy and Steve LeCheminant.

Left: C2 Class 0-8-0 035 is seen heading uphill towards the forests at Zhenzhu, six kilometres out from Weihe. The end of the line is almost 80 kilometres away at Liushan. We never quite discovered what the lanterns signified. Alan, our guide for the second decade of our China visits, said they were the equivalent of Michelin stars for restaurants! Certainly just down the road in Pinglin, we had one of the best meals ever in China, with delicious cuttlefish imported from Russia via the Sea of Japan. For me, this image captures village life along many of the Chinese logging lines.
Photographic Details: Canon EOS1n, Fuji Provia, Canon 70–200 f2.8L lens, 1/125 at f11.

Above: The snows of March 2001 in the far north-east of China were the heaviest for several decades. On 8 March 2001, QJ Class 7017 approaches the summit on the climb away from Shuangyashan with the 14.40 service to Bafenchang, which the colliery operated to carry its miners. With a loading of 11 ex-China Rail coaches, this looks every bit a main line service, whereas it was actually a local workers shuttle. The snow was so deep either side of the embankment, that I sank in it up to my shoulders earlier in the day whilst trying to find a good location and had to be rescued by my colleagues. **Photographic Details: Mamiya 1000S, Fuji Provia, Sekkor 150mm f3.5 lens, 1/500 at f5.6.**

Above: Perhaps my favourite shot of the Weihe logging system is this view taken at 09.00 on 11 March 2001. I mention the time because the loaded log train was running several hours late, due to the failure of one of the C2 locomotives. Normally several log trains would return overnight to the log yards at Weihe, but thanks to the failure, it was possible to capture this loaded train in the light. I failed to identify the numbers of either locomotive, but was delighted with the picture! **Photographic Details: Mamiya 1000s, Fuji Provia, Sekkor 80mm f2.8 lens, 1/500 at f5.6.**

Left: At the top of this image is the snow shed, protecting the hump at the rebuilt Argentine Yard in Kansas City, USA. Taken on 20 April 2001, this shows various BNSF Railway locomotives on "trim" duties, which involved drawing out freight cars from the various classification tracks in order to build departures. The 60-track hump yard here was opened in 1997, replacing two former hump yards on the site and is one of the busiest yards in the USA. My journey there had been somewhat eventful. Having had my research fellow present the previous day in St. Louis and then get interviewed by the New York Times, I was in buoyant mood, so decided to get up at 03.00 and drive to Kansas City for a quick visit before flying home. The problem was I was driving at 85–90 mph along the deserted highway when a couple of state troopers pulled me over. Being English, I got out of the car (which is apparently the wrong thing to do!) and so they spread eagled me over the bonnet and all I could say was "Sorry, I'm English". The meeting ended with them sagely telling me "You have a nice day now"! **Photographic Details: Canon EOS1n, Fuji Provia, Canon 70–200mm f2.8L lens at 200mm, 1/250 at f8.**

2002

During 2002, hardly any images were captured in the UK, although two trips to China were made, as was another to a Plandampf in Germany with Vernon Murphy. Descriptions of each year are beginning to sound like a broken record!

Above: This year was a low ebb for me as regards photography and for once my records are patchy, therefore all I can say is that this shot of German Pacific locomotive 01 0509 was taken in February 2002 somewhere between Gera and Saalfeld in the former German Democratic Republic. Indeed, but for the advent of digital photography, it is highly likely that I would have boxes of uncatalogued slides in a cupboard somewhere and dozens of unprocessed black and white films. When I inherited Vernon's photographic collection in 2014, I discovered packets of negatives which hadn't been opened since the mid 1960s, so busy had Vernon been professionally. In 2002, I was heading the same way.

Top right: During the doubling of the last sections of the A11 in 2002, there were regular stone trains from Peak Forest to the Johnson sidings at Eccles Road in Norfolk. In June of that year, 66172 shunts at Wymondham, after the train arrived and before it was split to be taken back to Eccles Road. In 2021 the warehouses in the background were demolished, to make way for large scale house building. Another quirk of this image is that it was taken with a Voigtlander Bessa rangefinder camera, which I used to keep in my work briefcase. I don't think I had it more than a couple of years before I traded it in, but it was as close as I ever got to owning a Leica! **Photographic Details: Voightlander Bessa 2, 50mm f2.5 lens, 1/250 at f8.**

Right: In March 2002, Vernon and I decided to make a short trip to China which was eventually dubbed the "crocodile hunting trip", so named because of the many "crocodile" electric locomotives we saw at Benxi steelworks and the Fuxin opencast mines. Here, on 11 March 2002, in unseasonally warm weather, German-built Bo-Bo 1013 shunts in the steelworks at Benxi. At the time, the town had a population of 1.56 million, 80,000 of whom were employed in the steelworks. Whilst the works still had a steam fleet, most steam locomotives seemed to be redundant and stored on the depot. **Photographic Details: Canon EOS1n, Fuji Provia, Canon 50mm f1.4 lens, 1/500 at f5.6.**

1013
1013

Above: The highlight of our March 2002 China trip was undoubtedly photographing the narrow gauge Shanhetun Logging Railway. The fact that it ceased operating just weeks after our visit, never to reopen, made this particularly significant. Several previous visitors had commented how four or five trains would arrive under cover of darkness, making them very difficult to photograph. We were very fortunate however, because on 15 March 2002, a serious derailment led to the loaded log trains being several hours late, which meant that they arrived during the first light of the morning. China Railways C2 Class number 06 arrives at the main yard in Shanhetun, slipping on the slight upgrade into the sidings. That morning saw us catch another two delayed arrivals before the lines were clear for the empty trains to return to the forests. **Photographic Details: Canon EOS1n, Fuji Provia, Canon 70–200 f2.8L lens, 1/250 at f5.6–8.**

Below: The colliery system at Tongchuan, a couple of hours from the historic city of Xian, was unique in that it utilised JF Class 2-8-2 locomotives well into the 21st century. During our first visit there in December 2002 we had exceptional weather. There were five collieries in the area and by far the busiest was at Wangshiwa. The climb from Tongchuan to the summit at Baipiaogou was severe, as the line on the right illustrates. From here, trains of empties drifted down the grade to Wangshiwa Colliery, which was on a short branch off this main line. JF Class 2369 reaches the summit on the morning of 8 December 2002. The signalman is producing almost as much smoke as the locomotive! **Photographic Details: Mamiya 1000S, Fuji Provia, Sekkor 80mm f2.8 lens, 1/250 at f8.**

2003

2003 was a momentous year photographically, as it was the year I "went digital". I had seen our Chinese guide Li Weishu with his four megapixel Canon 1D, but was worried that the resolution would be far too low. When the Canon 1Ds was announced, with an unheard of 11.3 megapixels, I decided to take the plunge. The image included in this section at Brundall was taken alongside a medium format transparency taken on my Mamiya 1000s. I had them both professionally printed and was convinced that digital photography had arrived. I managed two trips to China in 2003, visits to Dab Doberan and Rugen in the former East Germany were the result of escaping for the weekend after teaching at a course in Hamburg, allowing me to catch some narrow gauge steam.

Above: My first China trip of 2003 was in February and was "pre-digital". During this wide-ranging trip which took in a second visit to Tongchuan and a journey to the far west to visit the DaGu Coal Railway in the Gobi Desert, we also spent three days on the Jitong line and the pick of the images from this part of the trip was this sunrise shot at Jinpeng West Viaduct. Just five minutes after the sun popped up on the horizon, this pair of QJ Class locomotives whistled up in Jinpeng station and the semaphore dropped – perfect timing – proving the old adage that "it is often better to be lucky than clever". **Photographic Details: Mamiya 1000S, Fuji Provia, Sekkor 150mm f3.5 lens, 1/500 at f5.6.**

Left: Ever since moving to Norwich in 1995, Brundall was a favourite location, especially for capturing summer Saturday workings. This shot on 5 April 2003 was the twentieth image taken on my new Canon EOS1Ds and the picture that convinced me I could get medium format quality on a digital camera. It is astonishing to think that the camera body cost £6000 and yet advances in digital camera technology mean that the quality of a sunny day portrait like this, can now almost be matched by an iphone. 150257 pauses at Brundall with a service for Great Yarmouth. **Photographic Details: Canon EOS1Ds, ISO 100, Canon 70–200mm f2.8L lens at 135mm, 1/400 at f8.**

99 2323-6
CITROEN-
FÜR ALLE FABRIKATE

Left: During the summer of 2003 I was invited to teach on a laparoscopic skills course in Hamburg. When this finished, I picked up a hire car and made a 48-hour trip to Bad Doberan, Rostock and the island of Ruegen. The delightful Mollibahn, based at Kulungsborn, runs for over a kilometre through the streets of Bad Doberan. It is possible to sit in a café with a wine or a beer and take pictures of the train on the main shopping street without the need to even get up! Time really has stood still here as a second visit in the summer of 2021 revealed that the shops remained the same and even the storefront displays were strikingly similar. My favourite view is however on the approach to town, where cars and train mingle without any fuss or fencing. Ornstein & Koppel 2-8-2T 99 2323-6 heads down Goethestrasse on 11 July 2003. **Photographic Details: Canon EOS1Ds, ISO 200, Canon 70-200mm f2.8L lens + 2x converter at 400mm, 1/250 at f6.3.**

Below left: This image from 2003 captures a commonly repeated scene from what has become over a quarter of a century of trips to China. We were sitting by the lineside in the winter sun, with a bottle of beer, when this track worker wandered past, waved and then carried on his way up the line towards Wangshiwa. Silence returned and we sat listening to the birds singing – something we must have spent thousands of hours doing, whilst waiting for the next train. **Photographic Details: Canon EOS1n, Fuji Provia, Canon 70-200 f2.8L lens at 200mm, 1/250 at f8.**

Below: Our second trip to China, in the autumn of 2003, took in the Hechi Railway in China's Guangxi Province in the deep south. The railway passes through some dramatic Karst mountain scenery, as seen here at Wenping just north of Hechi. On 19 October 2003, we spent a delightful couple of hours on the station here watching a litter of puppies chase round the bundles of soya beans that had been laid out to dry on the concrete surface of the station platform. In the autumn sun, the beans could be heard popping and dropping to the floor, ready for collection by the local farmer, who swept them up into bags. China Railways JS Class 2-8-2 8285 arrives at Wenping with the afternoon mixed train from Pingzhai to Hechi. The mines north of here were producing 900,000 tons of coal a year at the time of our visit, but it looks as if they have all closed now and the railway line is mostly abandoned. **Photographic Details: Canon EOS1Ds, ISO 200, Canon 70–200mm f2.8L lens at 70mm, 1/500 at f9.**

2004

A significant change in 2004 was a scheme by my employer at the time, offering additional holiday in exchange for undertaking extra operating lists, to help clear the waiting list. This was an attempt to reduce the large sums being given to staff to undertake lists at the weekend, on top of their routine work. Many surgeons were not interested in the idea, but I jumped at the chance and as a result, was able to make four trips to China that year. Again, I can only apologise to the dyed in the wool UK enthusiast, as my pictures in the UK were limited to fleeting moments, either whilst travelling to London with work, or travelling elsewhere in the UK to teach laparoscopic surgery.

Left: This picture had to be included for several reasons. It may look straightforward, but it took five visits over an eight-year period to capture the morning freight on Erdi Viaduct. This location is at the end of the unofficially named "Happy Valley", between Jinpeng and Shangdian, so called because the photographer could happily capture trains there all day, as the line turned through 180 degrees, allowing shots all day. One difficulty in achieving a picture here was that for 300 days of the year, it is windy on the Jinpeng Pass, so the steam blows around and is rarely vertical. In addition, the area is semi-desert and so snow is infrequent in winter; I rarely saw a good covering as here. Finally, there is only a two-hour window when this image is possible for eastbound trains, as the sunlight is only in the right place between 08.30 and 10.30. On 8 February 2004, Vernon and I got it all right, when QJ Class 6998 & 6981 crossed the viaduct. We had photographed the pair leaving Jinpeng station at 08.10, then chased them up the line, taking this view at 08.50. One final reason for including this image is that it won the British Airways Highlife magazine photographic competition a couple of years later – just about the only professional photograph competition I have ever entered and definitely the only one I have ever won. The prize was a Nikon digital camera, which wasn't much use to a Canon user! **Photographic Details: Canon EOS1Ds, ISO 200, Canon 70–200mm f2.8L lens at70mm, 1/500 at f10.**

Above: From our first visit to the Jitong line in January 1997 to my last in October 2005, we walked every inch of the route between Jinpeng in the west and Reshui in the east, as did many hundreds of other photographers from around the globe. It was therefore unusual to be able to capture an image that hasn't been reproduced a hundred times over, but this one I think is fairly original. The curves of the railway lines and road intermingle as a solitary QJ Class climbs over Simingyi viaduct. The view was taken from the trackside, to the south-west of Xiakengzi station, through which the train had passed about ten minutes earlier on 8 February 2004. **Photographic Details: Canon EOS 1Ds, ISO 200, Canon70–200 f2.8L lens at 200mm, 1/400 at f14.**

Above right: During 2004, I still carried a film camera with me, loaded with Ilford XP2 black and white film. On our February trip to China, the Canon EOS1n shutter button broke and I was left in the middle of Inner Mongolia with just my digital camera. The nearest large town was Hohot and I went camera shopping – a Seagull SLR was the only camera available anywhere in the city and was purchased for £50. It still works perfectly today. I am pictured (left) with this camera around my neck, on a slag heap in Nanpiao with the late Vernon Murphy, enjoying mid-morning coffees, and an orange and chocolate muffin. We look a right pair of Charlies!

Opposite page top: This is the first of two images of the delightful Shibanxi narrow gauge railway in China's Sichuan Province, where we made the second of our three visits in February 2004, accompanied by Nigel Harris and a couple of his colleagues from Rail magazine. The 20-kilometre line provided the only access to the town of Bagou and its 4000 inhabitants, there being no road that far up the valley in 2004. That gave the remarkable site of the daily passenger trains being used to transport anything too heavy to carry, including sofas, fridges and other household goods, not to mention livestock. Pigs for slaughter in the main town of Yuejin had a special pen in the middle carriage, next to where the school children sat! The pigs at Xianrenjiao clearly know something unpleasant lies ahead, as they are most reluctant to board the train. **Photographic Details: Canon EOS 1Ds, ISO 200, Canon 28–70 f2.8L lens at 45mm, 1/320 at f3.2.**

Below: The second image of the Shibanxi narrow gauge railway has a tale behind it. After an 05.30 dumpling breakfast, we spent a delightful morning walking from Bagou to Xianrenjiao station, where we spent several hours chatting (through our guide Zebedee) with the station master, his family and a local barefoot doctor, who had a small surgery and pharmacy at the station. Lunch approached and the station master's wife suggested that she cook us something, to which we agreed; the catch was that we had to catch the chicken we wanted, whilst she picked the vegetables from the small plots dotted around the station. She sent a friend out to a neighbour, to find us some beers and lunch was superb. As we were enjoying a post lunch snooze, this train, which was not in the timetable, appeared. The ladies were playing Mahjong and the toddlers were pottering about on the tracks as the C2 Class engine trundled to a stop. The barefoot doctor's surgery can be seen behind the locomotive. An angry Englishman leaped off the train and ran towards us shouting "You can't photograph this train, we've paid for it, you'll have to give us a tip". As quick as a flash, one of our party quipped, "I'll give you a tip, f*** off and be nice to your mother!" Eventually the gentleman calmed down and perhaps as a result of my companion's eminent position in the railway world, we ended up sharing a whisky with said gentleman that evening in our hotel in Qianwei. **Photographic Details: Canon EOS1Ds, ISO 200, Canon 28–70 f2.8L lens at 42mm, 1/250 at f5.6.**

Above: A 20-minute diversion off the A1M on 30 April 2004 found me lucky enough to capture a coal train from Harworth Colliery hauled by 66524. This is seen at Firbeck South Junction, as it winds out of the colliery sidings. The picture was taken with the newly released Canon Powershot Pro1, the forerunner of the popular Canon G Series cameras. I have to confess that I had much to learn about digital photography in those days and I was ignorant of the noise problem that smaller sensors gave when the sensitivity was pushed to 400 ISO or higher. At this stage, I found it amazing that one could have an 8-megapixel camera with built in 28–200mm zoom which even had the "L" branding denoting a professional lens with dispersion glass. Within a couple of years, the limitations of results when setting the ISO above 100 had become apparent and the camera was passed to one of my children, but not before I took several thousand images at home and abroad with what at the time was an excellent little camera. **Photographic Details: Canon Powershot Pro1 at 700mm, ISO 100, 1/250 at f4.**

Below: Having new responsibilities in London with the National Institute for Clinical Excellence and the Association of Laparoscopic Surgeons of GB & Ireland, I found myself on the train to town every one or two weeks. Life was so busy, that apart from during my trips to China, I have no photography notes for the period between late 2003 and August 2006, so I don't even know the numbers of these two Class 90s that are posing at Norwich on 8 October 2004; the grain from the small digital sensor prevents enlarging the picture to find out. **Photographic Details: Canon Power Shot Pro1 at 45mm, ISO 100, 1/200 at f7.1.**

2005

This was the second year where extra annual leave was on offer to help get more operating lists undertaken. As a result I managed four trips to China. This was particularly fortuitous, because steam on the Jinpeng line finished in December 2005 and I managed to visit the route on three occasions, in January, July and October 2005. Little did my employers realise that extra annual leave was exactly what I wanted in 2005! I also made several visits to the USA as an invited lecturer, but otherwise took only a few pictures close to my home in Norwich. 2005 was also the year I bought a second digital SLR in the shape of the Canon EOS 1Ds Mark II, which had a ground-breaking 16.7 megapixels. This was the camera that finally consigned all my remaining film cameras to the scrap heap.

Above: Like many photographs that turn out "just right", this often comes down to luck rather than good judgement. That was definitely the case for this shot of QJ Class 6988, which was taken at 05.10 on 23 July 2005. Steve LeCheminant and I had risen at 04.30 to find a dense mist. This had put our Australian colleagues off getting out of bed, but Steve and I headed west from Lindong, towards the viaduct at Chaganhada. We couldn't see it when we arrived, but the mist burnt off quickly as the sun became stronger and just as the viaduct became clear, we heard the whistle of a freight train, looped at Dariqiga – perfect timing. Needless to say, when we joined up with the Aussies later in the day, they rued staying in bed. One cardinal rule of photography – if you aren't out with your camera, you definitely won't get the shot! **Photographic Details: Canon EOS 1Ds, ISO 200, Canon 28–300mm f3.5–5.6L lens at 80mm, 1/400 at f9.**

Above: There are some locations that one returns to again and again and for me, the Norwood Road overbridge at March Whitemoor Yard is one of those. Having first taken pictures here in the mid-1970s, when Whitemoor Hump Yard was busy and wagonload freight abounded, I was grateful that Whitemoor was repurposed as an infrastructure yard and freight there continued. On 7 May 2005, 66705 leaves the yard with a train of ballast for East Anglia. The platelayers hut on the right looks exactly the same as it did in my 1970s pictures, whereas everything else has changed almost beyond recognition. The Wisbech line, which begins behind the eighth wagon, is still intact, but the mighty Whitemoor Junction signal box is long gone; the 147-lever box closed in 1987. The housing estate behind the train is new and the nearby locomotive depot has long been demolished. **Photographic Details: Canon EOS 1Ds Mk II, ISO 200, Canon 28–300mm f3.5–5.6L lens at 80mm, 1/400 at f9.**

Below right: I was invited to give some lectures on surgical research at New York University in Manhattan in April 2005. There is virtually no freight on Manhattan and the major freight yards for the area (and for many other large cities in the USA) are located many miles away in rural areas. There is however a small hump yard at Oak Island on the New Jersey side of the Hudson River. I had to investigate and found that I was able to get a shot of a freight train which included the Empire State building. It may sound a bit quirky, but I was pleased to include the iconic structure in a railway image, especially after the atrocity of 9/11. **Photographic Details: Canon EOS 1Ds Mk II, ISO 200, Canon 28–300 f3.5–5.6L lens at 270mm, 1/320 at f10.**

Above: Freight traffic around my hometown of Norwich was scarce compared to the other places I had lived, such as Newcastle and Cardiff, so to capture two non-passenger trains in the same frame was a rarity. On 15 December 2005, 60063 stands in the loops at Wymondham with the second portion of a Peak Forest to Norwich stone train. The reduced siding capacity at the remnants of Norwich's goods yard meant that limestone deliveries for the sugar factory at Cantley had to be split at Wymondham. 20304 and 20311 pass with the daily railhead treatment train. **Photographic Details: Canon EOS 1Ds Mk II, ISO 320, Canon 28–300mm f3.5–5.6 lens at 300mm, 1/320 at f6.3.**

Left: I made my last visit to the Jitong line in October 2005. It was marred by my luggage being lost in Amsterdam's Schipol Airport for a second time and as a result, I was frozen solid taking this image at dawn on 9 October 2005, when the temperature was minus 6 degrees. My down jacket and boots were left in the aircraft hold and it took KLM six days to get them back to me. They gave me about £30 compensation when I arrived in Beijing and I bought a couple of sweatshirts and a fleece, anticipating cold weather on the Jitong line. That was the last time I used KLM for long haul flights! This image was taken at 06.15 on 9 October 2005, at Yameniao, just east of Lindong. As had often been the case on trips to China, fortune seemed to be on our side. The sun peeped above the horizon just as the semaphore dropped to signal the approach of a Chanbuga to Daban freight. QJ Class engines 7105 & 6984 made a spectacular start to the day. **Photographic Details: Canon EOS 1Ds Mk II, ISO 200, Canon 28–300mm f3.5–5.6L lens, 1/400 at f9.**

2006

I think waiting lists must have gone off the government agenda in 2006 because I only made one visit to China, although in retrospect that may have had more to do with the end of steam on the Jitong line. Instead, I did enjoy a one-week autumn trip to the USA with Bob Avery and a couple of days out linked to surgical meetings in Scotland and the north of England. The year also hosted one of my craziest surgical escapades, with a day trip to Japan to perform the first laparoscopic gastric bands in the country. I flew from London to Tokyo's Narita Airport, was chauffeured to Haneda and flown on to Oita, where a hovercraft took me from the airport to the University Hospital. Dinner was a "puffer fish supper" and at 07.00 the next day I was in theatre. Finishing operating at 15.00, I flew back to Tokyo for a ten-minute appreciation of the cherry blossom and a night in the Imperial Hotel, before flying home from Narita to London Heathrow on 26 March 2006.

Below: On the morning of 26 March 2006, I found I had a couple of hours to spare before I needed to head to Tokyo Narita airport, so I wandered over to Tokyo Central station. I was able to capture dozens of trains during the rush hour, including this Nozomi Bullet Train. It was only my second visit to Japan and I was delighted to leave with at least a few railway pictures! **Photographic Details: Canon Power Shot Pro1 at 110mm, ISO 100, 1/250 at f4.**

Above: I have always liked this view in Logan, West Virginia, as it seems to capture "run-down middle America". Logan has one of the highest obesity rates in the USA, with virtually everybody classified as being morbidly obese and the food outlets reflect this. The square to the right of the train was used as the location for a Channel 4 News piece during the run up to the election of Donald Trump as President. It was startling to later recognise the spot from the comfort of my sofa, when it popped up on the news. Getting this shot on 26 October 2006 was far from easy. We had photographed this train up the branch line, at what I have recorded as Buffalo Creek Mine (Pardee). We waited for it to leave, as we were confident that we could chase it back to Logan, which was fine as far as Amherstdale, a few miles down the line. What we had not realised was that the line speed dramatically increased after Man Junction, with the train hurtling along at 50 mph from there. We had a knife edge crossing in front of it, halfway down to Logan and a frantic dash through the town to this location, which we had selected earlier. It is a loaded coal train from Pardee to Peach Creek Yard, just north of Logan. During September 2006, Peach Creek Yard handled a total of 10,112 cars of coal, which was a record at the time. **Photographic Details: Canon EOS 1Ds Mk II, ISO 400, Canon 70–200mm f2.8L lens at 105mm, 1/320 at f5.6.**

Left: This view, taken on 26 October 2006, made me chuckle. I had discovered that Halloween was a big event when I lived in Minnesota in 1967 and 1968, but had forgotten how seriously the Americans take it. At Amherstville, where a train of 150 cars was loading at the large bunkers out of sight round the corner, CSX Transportation 478 is one of the train's four locomotives, with two at the front and two at the back. The town was dressed to the nines in preparation for Halloween, with every house competing to have the best display. **Photographic Details: Canon EOS 1Ds Mk II, ISO 800, Canon 70–200 f2.8L lens at 70mm, 1/50 at f25.**

Below: A location I had long hoped to visit was the viaduct just south of Montrose station. The Association of Surgeons 2006 Spring meeting just happened to be held in Edinburgh and so on 4 May 2006, I hopped on an afternoon train from Haymarket and met Bob Avery at Montrose. Our visit was brief, as I needed to be back in Edinburgh by the evening, but we managed to catch four freight workings within two hours and take some images of the viaduct and the soon to be closed Usan signal box. A Scotrail Class 170 working an Aberdeen to Glasgow service is captured on the viaduct. **Photographic Details: Canon EOS 1Ds Mk II, ISO 320, Canon 28–300 f3.5–5.6L lens at 60mm, 1/400 at f11.**

Above: On 30 June 2006, I had a day out on the Settle and Carlisle line with Steve LeCheminant when we had perfect weather. 66170 is seen passing over Arten Gill Viaduct with a Hunterston to Drax MGR working. The day was so warm that when we had dinner with Tim Littler in Altrincham that night "al fresco", it was still balmy T-shirt weather at midnight – unheard of for Greater Manchester! **Photographic Details: Canon EOS 1Ds, ISO 320, Canon 28–300 f3.5-5.6L lens at 58mm, 1/500 at f8.**

2007

By 2007, the storm clouds were gathering on the professional front. The NHS I had joined in 1984 was changing incredibly rapidly. A colleague who retired in 2006 gave a parting lecture, pointing out that in 1976 there were eight managers in our hospital, but that number had increased to approximately 800! There was also a relentless reduction in the priority given to research, except in big international centres of excellence such as Cambridge or London, making it especially difficult if one's major professional interest was research. Managers increasingly wanted surgical automatons to just cut through the work and do as they were told, rather than free-thinking consultants who had national and international commitments. Looking back, it's easy to see this, but at the time I was simply the frog in the pan of lukewarm water, with the gas turned on low! On the photography front, I made two visits to China early in the year and was persuaded by Vernon Murphy to join him on two trips to pastures new in Zimbabwe and Argentina. Both were excellent. Travel with work remained frequent, with a weekend lecturing to the students and staff at the University of California San Diego, which allowed a quick afternoon flit to Tehachapi. Similarly, a visit to the annual laparoscopic meeting in Newcastle enabled me to visit a few old haunts.

Right: On 18 April 2007, after lecturing in San Diego, I had to travel to Las Vegas, where our team presented six research papers at the annual American Laparoscopic Meeting (SAGES). This was a perfect opportunity to revisit Tehachapi, but this time with a DeLorme Topographical Atlas of Southern California. This enabled me to finally find a vantage point over the Walong Spiral, which I understand is now a more formal "railfan" lookout. After clambering up several hills, this view seemed to be the best vantage point to capture the whole spiral. A BNSF Railway intermodal (trucks on flat cars, which I think is a Richmond to New York service, heads south. The freeway from which my 1993 picture was taken can be seen on the right. **Photographic Details: Canon EOS 400D, ISO 200, Canon 17–85mm f4–5.6 USM lens at 22mm, 1/500 at f8.**

Below: Fishermen line the quayside on the River Tyne at 15.30 on 21 November 2007. The area had been completely transformed since the 1980s, with the new Copthorne Hotel taking pride of place. The pubs had all been gentrified, but the magnificent Tyne bridges remained a constant. **Photographic Details: Canon Power Shot G9 at 40mm, ISO 100, 1/1250 at f4.**

Right: This is the end of the 750 mm (2 ft 5½ in) gauge line known as "La Trochita" in Patagonia, Argentina. We had travelled the length of the 250-mile route from Ing. Jacobacci station and our locomotive is seen at the terminus in Esquel. This photograph was taken from the top of the water tower there, because I was keen to include the snow-capped Andes in the background. Vernon and I joined a Trains Unlimited Tour which visited Bariloche, the Esquel Line and the Rio Gallegos Colliery Railway. Despite a few disappointments, we took some good pictures and made up for the lack of steam that we expected to see on this trip by drinking Malbec and eating delicious Argentinian steak burgers in the bar car!
Photographic Details: Canon EOS 1Ds Mk II, ISO 200, Canon 300mm f4L lens, 1/800 at f9.

Above: In July 2007, Vernon and I joined an excellent railway-themed tour in Zimbabwe that was organised by Geoff Cooke. On our private charter train, we had arranged to meet Adrian Roberts from Sydney and made the new acquaintance of Richard McEvoy and his band of merry men. The bar bill for the eight of us, after five days on Geoff's train, came to over $1000, based on $1 per bottle or can of beer. Even though drinks accounted for the vast majority of the bar bill, I was questioned about buying four rounds of cheese toasties at $1 per toastie!

This image won the unofficial photo contest run by Geoff and shows a Beyer Peacock-built Garrett passing over Lukosi Bridge on 22 July 2007. The train was a charter freight that Geoff had organised and we travelled in the coach on the rear which is just visible. As dusk fell, Geoff advised us to return to the train pronto, as apparently this is a well-known watering hole for the local wildlife, which included several prides of lions! **Photographic Details: Canon EOS400D, ISO 200, Canon 17–85mm f4–5.6 lens at 17mm, 1/250 at f10.**

Above: For me, the shot of the trip to Zimbabwe was this one at Victoria Falls. We had checked into the Victoria Falls Hotel the evening before. I was upgraded to a suite, I think because the staff thought I was a relative of Cecil Rhodes. After a supper which included alligator with lemon juice (a culinary first for me), we spent the day travelling on our train across the border to Zambia. The morning sun has created

a rainbow through the spray of the mighty Victoria Falls. Some of our party took a helicopter ride to photograph this train at extra cost, but I was happy with the pictures from terra firma. **Photographic Details: Canon EOS 1Ds Mk II, ISO 200, Canon 24–105mm f4L lens at 28mm, 1/400 at f7.1.**

2008

Apart from two trips to China, the highlight of the year by some margin was my six-month exhibition at the National Railway Museum (NRM). I suspect like most good things in life, this happened by chance, rather than design. I had been invited to join Tim Littler of Golden Eagle Luxury Trains, as both medical support and a guide on the January 2005 Jinpeng Orient Express. I don't know why, but I happened to bring some prints of my recent China trips, which I put in the bar car for participants to browse through. Two of the guests on the trip were Andrew Scott, Director of the NRM and Colin Divall, Professor of Railway History at York University. Over dinner it transpired that all the national museums in Britain had been charged with setting up a China themed exhibition for the 2008 Beijing Olympic year. A preserved Chinese KF Class 4-8-4 was on display in the NRM's Great Hall, but Andrew hadn't decided how to make this into a special event. He asked if I might be willing to exhibit some of my pictures alongside the engine. In the end, three locomotives were moved off the museum's turntable and 66 of my pictures were displayed in a six-month exhibition which opened on 19 March 2008. It was especially poignant for me, as my parents, two of my children and my partner Irene were all able to attend and we had an after-launch party in the gallery overlooking the great hall.

Two other notable outings in 2008 were a visit to Eritrea to photograph the amazing Asmara to Massawa line and a Phoenix Railway Photographic Circle trip to Barnetby and Scunthorpe. I am reminded that I was so exhausted in November 2008 that I fell asleep in the guard's van while travelling around Scunthorpe Steelworks and took virtually no photographs!

Above: The sun is just popping over the horizon at 08.10 on 11 December 2008, as schoolchildren rush in front of an approaching coal train at Didao Washery. By this time, the Jixi Colliery system, which we first visited in 1998, had become one of the most popular places to visit in China. This had over 30 SY Class 2-8-2s spread over five distinct colliery groupings, of which Didao was the third largest, with four SY locomotives allocated to service the three collieries and two washeries in the area. **Photographic Details: Canon EOS 450D, ISO 400, Canon 70–200 f4L lens at 75mm, 1/250 at f5.**

Above: One of the signature shots in Eritrea was at Devil's Gap, which is seen here in a patch of sun on 4 November 2008. The wind that day was very fast and unpredictable, making it difficult to catch the train in the sun when it shone between the clouds. Four run pasts were arranged, where the train was called by radio when a break in the clouds was anticipated. Hugh Ballantyne led our group and had at least half a dozen advisors positioned to call the train. The crew were getting worried as they were running low on water for the trip back up the hill to Asmara, but in the end the timing finally proved favourable and the train was caught in the sun. The locomotive is a 442 Series 0-4-4-0T. **Photographic Details: Canon EOS 1n, Fuji Provia, Canon 70–200 f2.8L at 200mm, 1/250 at f8.**

Above: From left to right: Maurice Goodin, Vernon Murphy, your author, an unidentified gentleman, Richard Macavoy, Andy Wells, Jim Kay and Adrian Roberts from Sydney. Standing in Asmara station, we are wearing T-shirts that were designed for the trip by Adrian Roberts. One of Andy Wells' famous sayings when taking photographs was "Faaan...tastic" only for him to look at the image on his Canon G9 screen and announce "It's all gone horribly wrong!" This was printed on the rear of the shirts which Adrian handed out at a rate of one or two a day until Andy became desperate and asked if there was one for him, which of course there was. Sadly, my T-shirt finally fell apart a couple of years ago. **Photographic Details: Image taken on Maurice Goodin's Nikon D80.**

Above: This night shot, taken at St. Pancras on 11 January 2008, continues the theme of "the best images are not always taken with the best cameras"! To this day, it is probably my favourite shot of my favourite passenger station in the world, yet it was taken with a "happy snappy"! **Photographic Details: Canon PowerShot G9 at 28mm, ISO 80, 0.3 seconds at f2.8, hand-held.**

Left: London St. Pancras was a favourite spot for Irene and I. We often took the train from our home in Attleborough to King's Cross and would have lunch preceded by champagne at the fantastic bar at St. Pancras. In the early days as the Eurostar terminus, the station employed a falconer to keep pigeons at bay and he is seen at work on 11 January 2008. We had no idea what type of bird he used, but we now know it was a Harris Hawk. In 2019 Irene took up falconry and now has her own Harris Hawk called Ruby, which we use to keep pests away from the vineyard. In 2008, we would never have guessed that we'd later have a hawk as part of the family! **Photographic Details: Canon G9 at 35mm, ISO400, 1/20 at f2.8. Inset: iphone 12 pro max.**

Above: Dafydd Whyles arranged for members of the Phoenix Railway Photographic Circle to have an official visit to the signal boxes at Barnetby East and Wrawby Junction on 22 November 2008. The signalman at Barnetby East was exceptionally good natured and tolerant as can be seen from his benevolent expression, while surrounded by (amongst others), Kim Fullbrook, David Pearce, Martyn Fordham, Ian Cowley and Steve Arthur. **Photographic Details: Canon EOS450D, ISO 400, Canon 18–55mm f2.8 lens at 55mm, 1/30 at f7.1.**

Below: In spite of the scrum in Barnetby East signal box on 22 November 2008, traffic ran faultlessly and we even had the bonus of parallel trains to the east of the box. 60100 approaches with an iron ore train that is travelling from Immingham Ore Terminal to Santon, while 185141 overtakes with a Cleethorpes to Manchester Airport service. **Photographic Details: Canon EOS450D, ISO 200, Canon70–200 f4L lens at 121mm, 1/400 at f6.3.**

Above: In September 2008, the Association of Upper GI Surgeons met in Liverpool and our team had some presentations to make there. I decided to fly from Norwich to Manchester and catch the train to Liverpool. As was the case for several years around this time, I travelled with just a large briefcase and my camera of choice that day was the Canon PowerShot G9. I grabbed this snap upon arrival in Liverpool, which from left to right shows a Class 350 EMU, three Class 150s and a Class 156/142 combination. **Photographic Details: Canon Power Shot G9 at 28mm, ISO 100, 0.5 seconds at f2.8, hand-held.**

Below: This image of a Class 390 "Pendolino" arriving at Manchester Piccadilly was also taken in September 2008, on the way back from the surgical conference in Liverpool. Photographically, I was beginning to realise just how much smaller sensors were improving and to see that having a large DSLR and a couple of lenses was no longer essential to obtaining publishable images. Since this image was taken, I have experimented with several pocket cameras, from the Canon G9 & 15, through to the larger one inch sensor Lumix and Sony models and in 2021 I use the Sony RX100 VII. One thing that manufacturers don't all seem to have understood is that a small custom leather case makes all the difference when carrying a small sensor camera in one's pocket. Both the Sony and the Canon cameras score highly in this regard. **Photographic Details: Canon G9 at 28mm, ISO 80, 1/250 at f5.**

Left and below: Between 19 and 21 March 2008, there were several events to mark the launch of my National Railway Museum China exhibition. The six-month exhibition was probably the pinnacle of my photography over the last 50 years and something I was delighted to be able to undertake with the help of Andrew Scott and his staff, most notably Bob Gwynne. From left to right, the main image below shows Andrew Scott, Vernon Murphy, myself and Graham Hubbard of Bachman who kindly sponsored the launch party, complete with drinks, Chinese dancers and calligraphers. A representative of the Chinese embassy introduced the exhibition and even though he was a bit negative about it depicteding Steam Engines (he wanted to project China as modernising, rather than still having steam trains!), he was very interested to talk to my mother and father, with family ties being foundational in Chinese culture. The two other images show Irene and I visiting the exhibition unofficially on the morning of 21 March 2008 and then dressed up for the formal tour and book signing that evening. **Photographic Details: Main image: Canon EOS 1Ds Mk II, ISO 200 (flash), Canon 24–105mm f4L lens at 24, 1/60 at f4. Both secondary images: Canon G9 at 28mm, ISO 100 (flash), 1/60 at f2.8.**

2009

Unusually, no trip to China was made during 2009 and all my travel that year was related to work, except for one outing on a private charter organised by the late Dave Rogers. This visited Germany, Austria and the Czech Republic, but I bailed out of this early because of pressures at work. One big day out in the year was to steam my gauge one coal-fired QJ at a railway in Freckenham in Cambridgeshire, which there is more about in the captions. My only two purely railway-related outings in the UK during 2009 were an August visit to the North Yorkshire Moors Railway (NYMR) to celebrate my friend Jim Kay's birthday and a 24-hour trip to try and record the late evening freights leaving Mossend Yard.

Jim Kay owns two Gresley Teak carriages, an observation saloon and a kitchen car, which until 2021 were used by the NYMR. He has been kind enough to invite Irene and I to his birthday bash for the last decade or so and always provides an exceptional spread. Sadly, his carriages recently moved away from the NYMR, so this marvellous weekend can no longer take place. 2009 was especially poignant because Irene and I had just got engaged and consequently we found ourselves being whisked onto the footplate of ex-LNER Class A4 "Sir Nigel Gresley". The Scottish trip, which took place close to the longest day of the year, was a lesson in how poor the performance of even the best DSLR cameras were in 2009, when one increased the ISO to 1600. The pictures from this outing are of the two marvellous stations at Glasgow Queen Street and Edinburgh Waverley, each with images from 30 years earlier for comparison.

Below: Sometimes an apparently mundane image assumes much more importance with the passage of time. This is one of the joys of railway photography, recording an everyday scene to the best of one's ability, to find out a decade or two later that the scene is of historic importance. Such was the case at Attleborough station on 14 April 2009, as the MOM (Network Rail Mobile Operations Manager) climbs one of the signals on his weekly rounds to replace the paraffin lamps. Shortly after this, the paraffin lamps were replaced with LED lamps and then at the end of 2012, the signals were replaced altogether. **Photographic Details: Canon G9 at 200mm, ISO 80, 1/640 at f5.6.**

Above: In October 2009, Vernon and I joined a Dave Rogers tour to Germany, Austria and the Czech Republic. As usual, the train runpasts were exceptionally well planned, but there was quite a crowd of photographers on the trip, making it difficult for us all to find an unobscured view. The freight only branch from Garching to the AlzChem factory in Bavaria, Germany had

some excellent locations where I found clear views. Prussian Class P 8 4-6-0 38 3156 is about half-way along the short branch line and is seen hauling a demonstration freight train which includes our carriages on 20 October 2009.
Photographic Details: Canon EOS 1Ds Mk II, ISO 400, Canon 24–105 f4L lens at 47mm, 1/400 at f10.

Above: On 2 July 2009, 170460 and 170452 stand under the roof of Glasgow Queen Street station. In 2009, virtually all passenger services from Queen Street to Inverness, Aberdeen, Edinburgh and Fort William were handled by DMUs. Contrast this to the view that I captured 31 years earlier on 8 July 1978, when 26010 awaits departure to Edinburgh and 27101 is at one end of an Edinburgh "push-pull" service. **Photographic Details: Canon EOS 1Ds Mk II, ISO 400, Canon 24 – 105 f4L lens at 58mm, 1/125 at f6.3.**

This page: On 8 August 2009, ex-LNER Class A4 60007 "Sir Nigel Gresley" waits in the loop at Goathland, with a Grosmont to Pickering train, the rear two carriages of which are hosting Jim Kay's birthday event. "Sir Nigel" played a significant part in the weekend's activities. On the Friday evening, it hauled us on the dining train and on the platform at Pickering Irene uttered the immortal words, "there is a big fire in there!" At this point, the crew whisked us onto the footplate for the run round at Pickering, much to the chagrin of several other diners. On the Saturday, I was fortunate enough to ride on the footplate from Goathland to Levisham, whilst after that excitement, Jim had prepared an excellent spread in his dining carriage. **Photographic Details (main image): Canon EOS 450D, ISO 400, Canon 17–55mm f2.8 lens at 17mm, 1/160 at f16.**

Above and below: On 27 September 2009, my Gauge One (1.75-inch) live steam QJ Class 2-10-0 engine had its last outing. Since then, I haven't had access to a Gauge One layout in order to run the beast. The locomotive came into being after I obtained the plans for a real QJ whilst visiting Changchun Works in December 1998. The plans were stuffed up my jumper at the time and a few years later I started to explore the possibility of having a QJ built. Tony, the engineer at the Romney Hythe & Dymchurch Railway at the time, was a renowned locomotive builder. He was fascinated by the sheer size of the QJ, which has a grate area four times that of a BR Standard Class 9F, and was commissioned to build the locomotive. Its size is well demonstrated, as it stands next to a Class A4 on the layout at Freckenham. Once up to pressure, the engine can run for a couple of hours, hauling as many freight wagons as can be found to put behind it. Hopefully one day it will steam again. One image was taken with a Canon Powershot G9 and the other with a Canon EOS 1Ds, but can the reader tell the difference? **Photographic Details: Canon Powershot G9 at 50mm, ISO 80, 1/30 at f4 and Canon EOS 1Ds Mk II, ISO 800, Canon 70-200mm f4L lens at 200mm, 1/125 at f5.**

Above and right: 67030 leaves Edinburgh Waverley and heads for the carriage sidings at Craigentinny, with the empty coaching stock from the previous night's Caledonian Sleeper service. A Class 90 and another Class 67 can be seen stabled in the bay platforms. Contrast this view at the east end of the magnificent station, with that from 30 years earlier on 4 September 1979, when 27112 and 27212 operated a "push-pull" service to Glasgow Queen Street, with one locomotive at each end of the train. **Photographic Details: Canon EOS 1Ds, ISO 200, Canon 24–105 f4L lens at 58mm, 1/400 at f6.3.**

The 2010s

2010

Apart from a single trip to China, acting as a guide to Richard McEvoy and his crew, not much happened in 2010. The reason was the beginning of a conflict at work and other than a welcome break to the North Yorkshire Moors Railway for Jim Kay's birthday bash, most of my time was spent trying to work out what was going on with work. One bonus was that the images taken in China were excellent.

During the first half of 2010 I was making weekly trips to London because of appointments to the National Institute for Clinical Excellence and the Association of Laparoscopic Surgeons. I always kept my trusty Canon G9 in my briefcase and on 4 January 2010 was treated to a perfect sunset when I arrived back at Ely station after travelling from London via Cambridge. **Photographic Details: Canon Power shot G9 at 200mm, ISO 80, 1/800 at f4.8.**

Above: In February 2010 I acted as a guide for some of those that I travelled with in Zimbabwe during 2007, when they made their first visit to China. I had warned everyone how cold it would be, but one member of the party brought just his ski jacket. It is rare to ski at minus 30 degrees and standing around waiting for trains at that temperature soon caused some consternation. During the shift change at Didao on 2 February 2010, the mercury dipped to minus 30. SY Class 0407 arrives with empties from China Rail, whilst 0950 & 1446 are serviced on the left. The inset photo gives an idea of how cold it was – two of the group had grown icicles within five minutes of leaving the bus! **Photographic Details: Canon EOS 1Ds Mk II, ISO 400, Canon 24–105 f4L lens at 105mm, 1/200 at f5.**

Below: During the shift change at Dongcheng Mine in Jixi, up to six SY Class locomotives would gather to be watered and for the crew change. On 1 February 2010, SY Class engines 0804 & 1351 sit in the main colliery whilst a third locomotive is watered alongside the main line. Just after this image was taken, we retired for a warming breakfast of scrambled egg with tomatoes and stir-fried aubergines in a restaurant adjacent to the level crossing from which this was taken. **Photographic Details: Canon EOS 1Ds Mk II, ISO 200, Canon 24–105 f4L lens at 73mm, 1/320 at f7.1.**

Above: The first two carriages behind Southern Railway Schools Class 30926 belong to Jim Kay and hosted his birthday bash on 14 August 2010. The generous provision of food and drink for three round trips between Grosmont and Pickering was too much for Irene and I, so we decided to skip the middle trip and walk to Goathland to burn off a few calories. 30926 is seen passing the shed at Grosmont with the regulator open for the climb to Goathland. **Photographic Details: Canon EOS7D, ISO 400, Canon 70–200mm f4L lens at 111mm, 1/400 at f10.**

Left: Our shed visit at Fuxin nearly didn't happen. Alan Fengwang, our guide, was forced to see three different "bosses" and part with 200 Chinese Yuan for us to enter the depot and workshops as the sun was setting. With just ten minutes of marvellous light left, SY Class 1460 is silhouetted in the workshop. Because the image was virtually monochrome, it has been made into black and white using Photoshop. **Photographic Details: Canon EOS 1Ds Mk II, ISO 400, Canon 24–105 f4L lens at 35mm, 1/60 at f11.**

2011

Perhaps the most significant photographic events of 2011 were precipitated by Steve Ashling's invitation towards the end of 2010, to record the last days of the mechanical signal boxes between Ely and Norwich, including that in my home town of Attleborough. The project spiralled out of control as more contacts around the country facilitated access to the remaining mechanical signal boxes on the network. Especially worthy of mention in this regard are Anthony McIntyre, John Illingworth, James Skoyles and Trevor Maxted. During 2011, I photographed a total of 115 signal boxes. As a spin off from all these visits, my UK photo count increased dramatically and I was able to take far more train images. With much more spare time during 2011, I also enjoyed a couple of Farrail trips to the Harzbahn and Namtu Mines in Burma. Irene and I were treated by Nick Dodson to a ride in the Queen Mother's former dining carriage on the Statesman Rail train. A remarkable event also took place early in the year when Deltic 55022 took charge of the Lynemouth alumina trains for a few weeks.

Left: Thoresby Colliery was one of the last deep mines to operate in the UK, not closing until 2015. It was unusual in that the foot crossing in front of the signal box there was a public right of way used by runners and dog walkers. On 18 April 2011 I was experimenting with the newly released Panasonic GH2, my first foray into micro four thirds format and mirrorless technology. This was used to capture 66709 on a West Burton to Thoresby Colliery empties working as it enters the exchange sidings for the colliery. **Photographic Details: Panasonic GH2, ISO 400, Panasonic 14–140mm f4–5.8 lens at 67mm (138mm in 35mm), 1/500 at f11.**

Left: GB Railfreight have been well known for their "enthusiast" senior management and their hiring policy for heritage locomotives. It could be argued that the sight of 55022 "ROYAL SCOTS GREY" trundling up and down the former Blyth & Tyne Railway's freight network in May 2011 was the most spectacular result of this. Here, just after sunrise on 13 May 2011, 55022 passes Marcheys House signal box with the first run of the day from Blyth Cambois to the Lynemouth smelter. **Photographic Details: Canon EOS 1Ds Mk II, ISO 400, Canon 24–105mm f4L lens at 70mm, 1/400 at f5.6.**

Below: During February 2011, Bernd Seiler ran a five-day series of charters on the Harzbahn in Germany, mimicking the trains of the DDR era. On 18 February 2011, Mallet 0-4-4-0T 99 5906-5 is on a charter freight between Gernrode and Harzgerode with a typical mixed consist from the 1970s. **Photographic Details: Canon EOS 1Ds Mk II, ISO 800, Canon 24–105mm f4L lens at 50mm, 1/160 at f4.**

Above: Bernd Seiler, the owner of Farrail Tours, is legendary for his meticulous planning and attempts to make things as authentic as possible. For this scene he hired and drove an old Wartburg car and arranged for two actors to dress in East German army uniforms. As one wag shouted "Bernd, your library books are overdue!" No. 99 7238-1 approaches the crossing with a Nordhausen to Wernigerode freight made up of five standard gauge wagons on narrow gauge platforms. This kind of authenticity doesn't come cheap and the tour was plagued by non-paying photographers trying to sneak up to the run pasts. Bernd got so stressed about this that he developed a severe oesophageal spasm and got a lump of bread completely stuck in his oesophagus leading me to have to take him to Nordhausen hospital for an endoscopy! **Photographic Details: Canon EOS 1Ds Mk II, ISO 400, Canon 24–105mm f4L lens at 55mm, 1/200 at f7.1.**

Left: On 31 August 2011 we were treated to a day out on the Statesman Rail Fellsman from Lancaster to Carlisle and back behind LMS Stanier 8F 48151. Mrs. Rhodes waves at the photographers' gallery as we cross Ribblehead Viaduct in the comfort of the Queen Mother's observation saloon. **Photographic Details: Panasonic GH2, ISO 400, 14–140mm f4–5.8 lens at 32mm (64mm in 35mm), 1/160 at f7.1.**

Below: The Burma Mines Railway opened in 1908 to serve the mines linked to a smelter at Namtu. Halfway along the route is the famous Wallah Gorge Spiral where the two-foot gauge railway loops over itself to gain altitude on the final approach to Tiger Camp. Kerr Stuart 0-4-2T No. 13, built in 1913, is at the bottom of the spiral, while 1928-built Bagnall 2-6-2 No. 42 is halfway round the loop. To aid adhesion, both locomotives have two workers on the front running board, to apply sand from the bags stored there. Whilst this pattern of operation may not be strictly prototypical it makes for interesting images. **Photographic Details: Canon EOS7D, ISO 200, Canon 17–55mm f2.8 lens at 42mm (67mm on 35mm), 1/400 at f6.3.**

Left: During the year I visited all the surviving mechanical boxes in East Anglia, which gave me time to spend at the lineside. One of the views I had been meaning to attempt for years was this, including Ely Cathedral. Fortunately, on 8 April 2011 a two-car Class 158 just fits between the trackside bushes here. **Photographic Details: Canon EOS7D, ISO 200, Canon 70–200mm f4L lens at 116mm (185mm on 35mm), 1/500 at f10.**

Right: This classic view at Barnetby on 2 June 2011 shows 60074 with an Immingham Ore Terminal to Santon iron ore working. Barnetby East signal box closed on 30 December 2015. Opened in 1914 by the Great Central Railway, when it built the route to Immingham Docks, the box had 72 levers. **Photographic Details: Canon EOS 1Ds, ISO 400, Canon 24-105mm f4L lens at 105mm, 1/500 at f9.**

Below: A freight branch which I had wanted to photograph for several years was the Boston Docks line. On 3 August 2011, Harry Needle Railroad Company's 08507 shunts loaded steel wagons onto the quayside with St. Botolph's Church, which is known as the Boston Stump, visible in the background. **Photographic Details: Canon EOS 1Ds Mk II, ISO 400, Canon 24-105mm f4L lens at 73mm, 1/500 at f13.**

2012

This year saw me resign from my consultant post in Norwich and spend a very enjoyable six months working as a consultant in Hull. This had the bonus of a new area to explore in detail and also plenty of heritage signal boxes to visit. My notes record that another 109 signal boxes were photographed during 2012. I was also granted security clearance to photograph Her Majesty the Queen as she boarded the Royal Train at the disused Middleton Towers station. I think I was the only photographer at the station and armed police were chasing other onlookers away from the neighbouring fields, so images of that outing are included in my selection for the year simply because of their rarity value. I guided a second trip to China and finally managed an official visit to the world's largest mechanical signal box at Severn Bridge Junction in Shrewsbury.

Above: Driving up and down between Attleborough and Hull for six months provided plenty of opportunities to stop on sunny days and try new locations. The windmill at North Leverton, after harvest and when the hedges have been trimmed, made an excellent spot for coal trains to and from Cottam Power Station. In the background is West Burton Power Station as 66149 heads towards Retford with 4R23, an empty coal train from Cottam to Immingham Docks. Cottam Power Station was decommissioned in September 2019 and there are plans to build nearly 2000 homes on the 650-acre site. West Burton Power Station is scheduled for closure in September 2022. **Photographic Details: Canon EOS7D, ISO 400, Canon 70–200 f4L lens at 166mm (266mm on 35mm), 1/500 at f11.**

Above: When I discovered it in 2012, using a stepladder to look over the high parapet on Monkmoor Road in Shrewsbury, I had never seen this view before. It allows inclusion of both the 180-lever Severn Bridge Junction signal box and also the 93-lever Abbey Foregate box. On 13 June 2012, 66529 provides the added bonus of a rare freight working on the avoiding line to the south of Shrewsbury station. It is hauling imported coal from Portbury Docks bound for Rugeley B Power Station. This flow was short lived and Rugeley B was decommissioned in 2016, with demolition of the cooling towers taking place in June 2021. **Photographic Details: Canon EOS7D, ISO 400, Canon 70 – 300 f4–5.6L lens 182mm (291mm on 35mm), 1/400 at f7.1.**

Above: My late mother loved railways and in particular steam engines, so when we found that a steam special was passing Attleborough, I took my mum and dad to the station. On duty at Attleborough signal box on 23 June 2012 was David Hilton, himself an enthusiast who later retired to work at Pickering on the North Yorkshire Moors Railway. He immediately invited us in and my mother was fascinated by the workings of the signal box. On one occasion at 5.45 in the morning during the winter of 2012, David used the signal box PA to invite me into the box for a cup of tea, as the first train to Cambridge was running late and it was minus six degrees outside. The box closed on 3 December 2012 and no more was there a warm welcome at the station. **Photographic Details: Canon EOS 5D Mk III, ISO 400, Canon 24–105 f4L lens at 24mm, 1/160 at f4.**

Right: I call this image "Sooty and Sweep". In the gloom under the blast furnaces in Beitai in northern China, these two were sweeping up with brushes made from twigs. If it wasn't so funny, it would have been tragic, as they moved the piles of dust from A to B and back again. It was quite a scramble to try and include them in an image with the SY Class locomotive that was shunting under the furnaces and this was my best attempt. **Photographic Details: Canon EOS7D, ISO 400, Canon 70–200 f4L lens at 135mm (216mm on 35mm), 1/160 at f4. Converted to black & white using Photoshop.**

Right: The whimsical expression of the resident signalman at Immingham East made this image for me. He is chatting with James Skoyles, as James and I enjoyed a late summer evening visit to the signal box on 18 July 2012. The box was to close later that year, on 2 December 2012. It was equipped with a 60-lever electro-pneumatic frame, the power for which was generated by the device seen above the signaller's head. The box was replaced by a modular-type structure a couple of hundred yards away, which had the incongruous installation of the single line token equipment for the light railway to Grimsby that was salvaged from the box, alongside a modern signalling panel. **Photographic Details: Canon EOS 5D Mk III, ISO 400, Canon 24–105 f4L lens at 40mm, 1/25 at f4.**

Below: Morants curve is an iconic location near to Lake Louise in the Canadian Rockies, which was named after the official Canadian Pacific photographer who took publicity images there. I had first visited Lake Louise in the summer of 1968 and my only recollection of this is pony trekking on a horse called "Little Brown". I had promised to take my wife back to the Rockies one day and so it was that we came to be here on 1 September 2012. The problem was that there were absolutely no trains for four hours. The weather closed in and we had a snow flurry, after which a rainbow appeared. Irene was definitely on the fed-up end of being fed up and so I decided to give up and just take a picture of the rainbow. As I got in the car to return to Banff, I heard a diesel horn in the distance and sure enough, this train appeared. I tried to remember where I had stood to capture the rainbow and approximated the shot with a much blander sky. Back at our hotel there was a photography shop in the lobby and the girl running it told me she had done a degree in photography and asked to see my images from the day out. She offered to show me how I could merge the rainbow with my train picture and this is the result. I certainly don't have her skills, but am amazed by the result she has created. So much so that I bought a tripod from the shop which I still use today! **Photographic Details: Canon EOS 5D Mk III, ISO 400, 24–105mm f4 lens at 40mm, 1/500 at f4.**

Right & below: On 12 June 2012, a unique event in the history of the Royal Train took place. The Queen was touring the country as part of her Diamond Jubilee. Because of the duration of the tour, the train was longer than usual and consequently, it would not fit into the platform at King's Lynn without fouling the points in the station throat. West Anglia Operations Manager, Steve Ashling, wondered if the sand sidings at Middleton Towers might be used and Sibelco, the owners of the quarry, enthusiastically agreed, sprucing up the long-disused station. Steve felt the event should be recorded for posterity, but security was tight. I was given security clearance by MI5 and even met and chatted to the Queen's very own "Commander Bond" who travelled with her on the train. He understandably insisted on not being photographed. In the main image, West Anglia's Steve Ashling chats to Andrew Frost at the crossing gates at Middleton Towers, as 67026 & 67006 wait to pull up to the platform. Whilst standing to the west of the station, the coaches were loaded with boxes and boxes containing the Queen's wardrobe and food for the trip. The smaller image shows Her Majesty chuckling with her lady in waiting as they alight from their car to climb onto the platform at Middleton Towers. I could hear her saying words along the line of "where are we going, this is interesting." Apparently, the royal household liked the Middleton Towers arrangement, because the former station is so close to Sandringham, but I don't think it has been used since for the Royal Train, making this a unique event. **Photographic Details: Canon EOS7D, ISO 800, Canon 70–300mm f4–5.6L lens at 140mm (224mm on 35mm), 1/250 at f9.**

Right: With the closure of the manual 191-lever signal box at Melbourne Spencer Street in 2011, Shrewsbury's 180-lever Severn Bridge Junction signal box became the largest mechanical signal box in the world. On 13 June 2012, I was fortunate enough to be allowed an official visit to photograph inside this box. Amazingly, there used to be quite a number of lever boxes with over 200 levers back in the 1960s, the largest of which was a 299-lever box in York. Severn Bridge Junction signal box is a listed building and although only half the levers remain active, it still controls 300 trains every 24 hours. **Photographic Details: Canon EOS 5D Mk III, ISO 400, Canon 24–105 f4L lens at 24mm, 1/50 at f14.**

2013

This was another quiet year workwise, after a short spell working in the civil service as Medical Director for the Medicines and Healthcare products Regulatory Agency (Devices) which turned out to be like a daily episode of "Yes Prime Minister". One good thing about the appointment was the opportunity to live within St. Pancras station itself, waking each morning to a view from behind the clock there, which was certainly compensation enough for the torrid time spent as a civil servant! Otherwise, 2013 was a peaceful year with plenty of time to get out and take photographs. Again, signal boxes played an important role in the year's photography, with 93 new boxes visited and captured on camera. I made one visit to China to photograph the marvellous Rongshan Prison Railway, which turned out to be very fortuitous because it had only opened to foreigners a few months before our visit and then just weeks after our visit, on 20 April 2013, the Ya'an earthquake hit, badly disrupting the track bed, leading to the closure of this charming narrow gauge route. I joined Farrail for a trip to Brazil and took my wife on holiday to the USA where we "accidentally" came across quite a few trains in the Pacific North West of the country. Finally, A4 steam locomotives played a part in the year, with the chance to photograph the great gathering of the surviving six A4s in November and an exhilarating ride behind "Bittern" during December.

Below: Not a bad view to wake up to! On 3 February 2013, at 07.30 in the morning, all six of the international departure platforms at St. Pancras are occupied with Eurostar services. The building is Grade I listed and the flats within it have large, very thin sash windows, with added insulation for those that face north. The flats facing Euston Road to the south however, are unbearably noisy, with the sirens of ambulances and police cars disturbing residents' sleep through the night and in the summer, the temperature can reach more than 40 degrees due to the windows facing south. **Photographic Details: Canon EOS 5D Mk III, ISO 400, Canon 24–105mm f4L lens at 24mm, 0.5 seconds at f7 (hand-held).**

Right: The classic shot at the lagoon south of Imbituba was a priority for our Brazil trip and Bernd Seiler had planned the event meticulously. Unfortunately, because of service disruption caused by the stalling of No. 205 at Rio Forita on 11 July 2013, the wrong set of wagons ran to Imbituba. The wagons were modern wagons with a white stripe along their sides. This led to a manic couple of hours in the sidings at Imbituba, where 20 railway enthusiasts from all over the world applied brown paint to cover the white stripes on the right hand side of the train, to try and remedy the situation! No. 205 is seen passing the lagoon, just minutes before the sun was off the line and the dark brown paint can clearly be seen on the wagons. I think it was worth the effort, but I doubt many casual observers realise how much "we suffer for our art"! **Photographic Details: Canon EOS 7D, ISO 400, 70–300 f4–5.6L lens at 89mm (142mm on 35mm), 1/500 at f10.**

Above: At the southern extremity of the Dona Santa Teresa Colliery Railway in Brazil, No. 205, a Skoda 2-10-2, passes the remains of a coal loading facility on 11 July 2013. It was hauling our empty coal train to Rio Forita at the end of the line; although the train was a charter, it contained wagons which were loaded with coal bound for the power plant at Tubarao. The plan had been to load the same number of wagons as the locomotive had handled back in the 1980s, when it first arrived from Argentina. Sadly, it stalled on the climb out of Rio Forita Mine, causing chaos on the network and blocking access to the colliery. The railway here was built by the British back in 1884 and continues to operate in 2021, albeit no longer with any operational steam locomotives. **Photographic Details: Canon EOS 5D Mk III, ISO 400, Canon 70–300 f4–5.6L lens at 70mm, 1/500 at f11.**

Above: Whilst many scenes recreated by Bernd Seiler of Farrail are tightly controlled, this one at Rongshan in China was not. We visited the railway in March 2013, when it was operating normally, carrying coal from two mines which had previously been prisons for political prisoners. The local government had only recently allowed foreigners into the area; indeed, many attempts to visit the line had ended in tears and it was only when a Japanese enthusiast dressed scruffily, pretended to be a local and took a few pictures, that the existence of the line was even confirmed. Western enthusiasts trying to visit the area stood out like sore thumbs. It transpired that the prison had closed, so it was deemed acceptable for enthusiasts to visit the area. On 20 March 2013, the morning departure from Rongshan to the mines at Shanziba and Yujiabian leaves the yard at Rongshan, just as the sun peeps over the horizon. Both mines were still heavily fortified, making for great photographic opportunities during our three-day visit to the system.
Photographic Details: Canon EOS 5D Mk III, ISO 400, Canon 24–105mm f4L lens at 35mm, 1/320 at f8.

Right: Even though most youngsters had mobile phones and designer clothes, they had never seen Westerners before, because the ban on tourists visiting the area had only recently ended. This led to us being mobbed everywhere by school children who wanted a selfie with us. This group took multiple snaps with their phones and I lent my camera to Ralph Montague to capture the madness of it all.

Below: I have included this shot because it is one of the first on which I used the newly introduced HDR option on Canon digital cameras. The camera takes three consecutive images, bracketing either side of the average light reading and then merges them. Unfortunately, unlike the system on Olympus cameras, any movement caused significant "ghosting" as the frames weren't matched up. On 22 March 2013, China Railways C2 Class 210 is seen shunting in the yard at Rongshan, whilst this elderly lady enjoys a late lunch. It took about 12 attempts to capture her without too much hand movement and even then, I later had to disguise the movement of her chopstick hand around the knuckles using Photoshop. **Photographic Details: Canon EOS 5D Mk III, ISO 400, Canon 24–105 f4L lens at 24mm, 1/25 at f11 (HDR).**

Left: During the summer of 2013, Irene and I holidayed in the Pacific North West of the USA. Staying for a few days at The Sandpoint Hotel on Lake Pend Oreille, Irene commented wryly that there seemed to be a very busy railway line in the background! Worse was to come for Irene when on 3 September 2013, we wandered along the beach in Sandpoint and I commented that it was a shame there weren't any bathers on the pontoon. This overlooked the railway causeway which was known as "The Funnel", because all traffic in and out of the North West on BNSF must pass through it. I persuaded Irene to swim out and then tried to orchestrate the pose I wanted, shouting over a distance of more than 100 metres. Let's just say, I won't make a film director and after several attempts, this was the best we managed between us. "Come in No. 2 your time is up!". The approaching speed boat is ironically much more important to the image than the receding Union Pacific double stack train. Irene was cold when she swam back and my experiment cost me lunch in Trinity's restaurant on the beach, with a bottle of Cloudy Bay sauvignon blanc thrown in. **Photographic Details: Canon EOS 7D, ISO 400, Canon 70–300mm f4–5.6L lens at 135mm (216mm on 35mm), 1/320 at f18.**

Right: Not every railway photograph has to be taken with expensive camera gear as this image taken on 5 December 2013 proves! The story behind the picture was an invitation from my late friend Vernon Murphy to join him on a trip from York to Newcastle, travelling behind A4 Class 4464 "BITTERN". The trip was to include 90 mph running, but started inauspiciously as gales the night before led to a sedate 50 mph run from York to Newcastle. Once in Newcastle, we joined Tim Littler, Graham Bunker, Tony Streeter, the Steam Railway Magazine team and several others for a superb lunch at Hotel du Vin on the Quayside. In fact, it was so superb that we nearly missed the train home. There had been flooding the previous night due to the aforementioned storm and our taxi arrived back at Newcastle Central less than 5 minutes before the train's departure. Vernon's arthritis caused him to struggle to board, managing it 30 seconds before the train started moving. I ran ahead to take this image with a Canon pocket camera and it was used on the cover of Steam Railway Magazine. We reached somewhere around 95mph on the trip home and enjoyed a party atmosphere all the way back to York! **Photographic details: Canon G15, lens at 11.2mm (63mm on 35mm), 1/60 at f1.8, hand held.**

Left: Vernon Murphy and I ate mushy peas and a meat pie in the presence of the last surviving LNER Class A4 locomotives. Dubbed "The Great Gathering" it was probably the most successful event the National Railway Museum has ever staged. Opportunities for decent photographs were limited and this was the best I could do on the night when we were strictly corralled within the Great Hall. Sadly, both 60008 "DWIGHT D EISENHOWER" and 60010 "DOMINION OF CANADA" have both since returned across the Atlantic to relative ignominy. **Photographic Details: Canon EOS 5D Mk III, ISO 800, Canon 24-205mm f4L lens at 24mm, 1/5 second at f4 (hand held).**

2014

Most of my surgical work in 2014 took place overseas in mission hospitals in India and Thailand, which was a great privilege and very humbling. With the exception of discovering the last remaining metre-gauge hump yard in Asia, north of Bangkok, which had been reduced to a pile of rubble by the time I visited, there wasn't time or opportunity for much in the way of railway photography on these trips. My notes record that another 77 signal boxes in the UK were visited. Featured in this snapshot of the year's endeavours is the wonderful box at Great Rocks Junction, which is amongst my favourite heritage signal boxes in the country, not least because of the warm welcome offered by one of its residents, Alex Fisher. One freight location that had eluded me until 2014 was Aberdeen Waterloo Goods Depot and it was great therefore to visit this whilst attending the Association of Laparoscopic Surgeons annual scientific meeting in November of 2014. It was in the summer of 2014 that my dear friend and fellow photographer, Vernon Murphy passed away at the far too young age of 69.

These two pages: Great Rocks Junction signal box has always held a special place in my memories. I first went there in 1983 and took the cover shot there for the OPC album "British Rail in Camera". I visited again in 1990 and the signalman invited me and my three-year-old son Jonathan in and even fixed for me to ride on a limestone train the following evening, when I travelled from there to Lostock Junction and back. I hadn't been to the box in ages when out of the blue in 2013, I was contacted through the photographic sharing site Flickr by Alex Fisher, a resident signalman there. He requested images for a book he was writing about Wath (he is still writing it!) and I suggested I visited in person. Many sybaritic visits followed in rain and shine and much ribald conversation was had. The image of Alex cooking breakfast for us on 18 June 2014 clearly shows a photocopy of my 1986 book "British Rail in Camera" pinned to the wall behind the frame. He had been influenced by some of my earlier books and was also an immense help researching archives for my 2016 Marshalling Yard book. The morning of the infamous "everything included" breakfast, there was a four-hour possession on the line for track repairs and we sat certain in the knowledge that there were no trains. Sadly, there have been few visits of late but hopefully there may be more in the future.

Right: A shot I had wanted to try for some time. 60010 begins to move its 2000 ton limestone train, the 6F05 Tunstead–Oakleigh, and crawls past Great Rocks Junction on 30 April 2014. **Photographic details: Canon EOS 7D, ISO 400, Canon 24–105 f4L lens, 1/250 at f9 with fill in on camera flash.**

Above: High above the loading point at Tunstead Quarry, 60063 is seen loading the lunchtime trip to Briggs Siding. The train will draw forwards to the loops north of Great Rocks Junction, where it will run round for the trip to Buxton. After a second run round it will head to Hindlow. **Photographic Details: Canon EOS 5D Mk III, ISO 400, 24–105 f4L lens, 1/500 at f11.**

Left: The symmetry of 22 wagons crossing the junction at Stainforth and Hatfield is pleasing as they catch the evening sun. 66043 hauls an Immingham to Drax biomass train formed of brand new wagons on 23 June 2014. **Photographic Details: Canon EOS 7D, ISO 400, 70–300 f4–5.6L lens at 214mm (35mm equivalent 352mm), 1/320 at f11.**

Left: This picture was possible thanks to Steve Ashling and Andrew Frost. Steve was the Network Rail Operations Manager for West Anglia and helped facilitate me visiting all the signal boxes in the region to record that dying breed, the Victorian mechanical signal box. Andrew was the Local Operations Manager and accompanied me to the new control tower at the rebuilt March Whitemoor Yard. Having visited this site many times when it was a hump yard in the 1970s, it was fascinating to see this view of the southern half of the old hump yard given over to Network Rail traffic. I selected this image for the book because of the wonderful light on 6 May 2014, when 09002 and 08925 pause between shunting. **Photographic details: Canon EOS 5D Mk III, ISO 400, 24–105 f4L lens, 1/400 at f14.**

Below: My part time working hours provided me with more time for photography, however, not wanting to miss any opportunities that may arise, I always carried a small camera when travelling to surgical meetings. The 2014 Association of Laparoscopic Surgeons was held in Aberdeen and I was delighted to get an early morning opportunity to record the thrice weekly freight to Waterloo Goods Depot, doubly so since it was a freight branch that had eluded me thus far. In gloomy conditions, on 27 November 2014, 66085 shunts china clay slurry tanks in the depot, having arrived with the 06.40 from Mossend. **Photographic Details: Canon G15 at 15.9mm (35mm equivalent 90mm), ISO 400, 1/125 at f2.8.**

Above: There are many layers to this picture of 66732 passing Sleaford North on 9 May 2014, while working the Middleton Towers to Barnby Dunn sand train. The signal box had been taken out of use a few weeks earlier, as part of the transfer of control of the line between Lincoln and Peterborough to the Lincoln signalling centre variously in April, August and November of that year. The view also shows the reinstated Sleaford avoiding line, which along with the 2021 remodelling of Werrington Junction, represents massive investment to allow freight to be diverted away from the East Coast Main Line and along the former Joint Line to the east. **Photographic Details: Canon EOS 7D, ISO 400, 70–200mm f4L lens at 112mm (35mm equivalent 179mm), 1/400 at f11.**

Below: This image, which was my first of a Class 68, shows 68005 passing Perth station with the daily Inverness to Mossend Tesco train. The day was also noteworthy as a speeding cyclist appeared on the bridge to the south of Perth station, armed with a camera and perpetual motion! This turned out to be Guy Houston and several years later we would discover a shared interest in German freight operations. Mine stemmed back to my visits to the country in the 1970s and Guy's from several years working there during the 2000s. This led to a collaboration, writing the new "Freight Only – Germany" series. Again, it is amazing how chance meetings can lead to long term friendships and projects. Another reflection on this image is the rationalisation in Perth; at the time of my first visit in 1976 there was an automated hump yard and lines of locomotives stabled along the tracks where the grey fuelling shed now stands empty. The passenger station retains its faded glory to a large extent. **Photographic Details: Canon EOS 7D, ISO 400, Canon 70–300 f4–5.6L lens at 300mm (35mm equivalent 480mm), 1/320 at f7.1.**

Above: Throughout our time living in Attleborough from 2008 to 2018, Ely was an important station for us. Often having to travel to London for work, I frequently found myself changing trains at Ely, which was just a 40-minute hop from Attleborough and the home of one of England's finest "English" restaurants, The Old Fire Engine House. Here on 13 December 2014, after a fantastic Christmas lunch at "The Old Fire Engine", the Attleborough train arrives on Platform 1 to take two slightly merry travellers home! **Photographic Details: Canon G15 at 30.5mm (35mm equivalent 140mm), ISO 80, 1/400 at f5.0.**

Left: On 29 September 2014, a wonderful opportunity was arranged by Trevor Maxted to visit three signal boxes. Stafford Nos. 4 & 5 and Liverpool Lime Street. The two Stafford boxes closed in 2015, with the box at Lime Street following on 12 July 2018. With 95 levers, it was Britain's largest surviving electro-mechanical miniature lever frame. It was an unappealing box like structure with a concrete roof designed to protect against bombing in WWII, even though it was not opened until 1948. Signalman John Piercy is captured nonchalantly propping up the frame as he refreshes his knowledge in the box on the day of our visit in 2014. **Photographic Details: Canon EOS 5D Mk III, ISO 400, Canon 24–105 f4L lens at 65mm, 1/8 second at f5.6.**

2015

The big event for UK railways in 2015 was the abolition of the mechanical signalling between Scunthorpe and Cleethorpes, which included the iconic structures at Barnetby East and Wrawby Junction. For me the year brought a new challenge working as a consultant surgeon in Taunton, helping to clear the waiting lists. It soon became clear that driving from Norfolk was a non-starter, but flying via Manchester to Exeter was quite doable. A year later, the direct flight from Norwich to Exeter with Flybe made the commute exceptionally easy. In terms of overseas trips, I joined Telerail on a trip to Switzerland, using this as an opportunity to try my new camera. I also made a crazy weekend visit to China, made possible by using some airmiles that would otherwise go to waste! My notebook for the year records that I visited another 56 new signal boxes and recorded them for posterity. Photographically a momentous change came in March as I abandoned Canon after 37 years. The reasons behind this decision were multiple, but prominent amongst them were the weight and bulk of the gear. A chance visit to Campkins Cameras on Kings Parade in Cambridge, late in 2014, sparked my interest in the newly released Olympus EM1. Several large format prints really impressed me, both the resolution and the sharpness of the lenses. I was also frustrated with the HDR option on the new Canon EOS 5D Mk III which seemed to have been introduced before it was ready. In addition, other manufacturers seemed to have stolen a march on Canon compared to the heady days of 2004 when they ruled the world with their 16.7 megapixel EOS 1Ds camera. So it was time for change and a new EM1 was purchased together with the 12–40 and 40–150mm f2.8 zoom lenses, equivalent to 24–80mm and 80–300mm on 35mm.

Below: Having worked in Hull in 2012, I kept in contact with several people there and so on 17 April 2015, I drove up to try and catch the GB Railfreight Class 47 working to Hull Docks. To my surprise there had been significant tree clearance around Melton Lane signal box, making the shot from the bridge to the east much more attractive. It was said that the clearance was in preparation for electrification, but with severe delays on all Network Rail related projects, nothing as yet has happened with this. 47812 passes with a Doncaster Down Decoy to Hull Docks gypsum working. **Photographic Details: Olympus EM1, ISO 200, Olympus 40–150mm f2.8 lens (plus 1.4 teleconverter) at 210mm (420mm on 35mm), 1/500 at f8.**

Above: On 10 June 2015 I enjoyed a day out on the Wherry lines with Steve Arthur and Martyn Fordham, both of whom were using the latest Canon gear and large L lenses. 47805 (front) & 47813 head away from the photographer with a Norwich to Lowestoft service, passing Reedham Junction signal box. When we returned to Attleborough, I downloaded three files of very similar shots, printed them out on A3 paper and laid them out for us to try and guess who had taken which shot – we hadn't a clue. After seeing the quality of the Olympus lens and its superior depth of field, with all the bricks on the bridge in the background sharp, as compared to full frame images where they were slightly out of focus, all my Canon gear was listed on ebay over the next 12 months. Hence this image deserves inclusion in the 2015 pictures. **Photographic Details: Olympus EM1, ISO 200, Olympus 40–150mm lens at 97mm (194mm on 35mm), 1/500 at f6.3.**

Below: Whilst working in Taunton I occasionally got afternoons off and used these to revisit South Wales a couple of times. On 30 June 2015, it was a perfect evening as 60092 passed Ferryside with the Theale to Robeston empty oil train. The image was taken using the excellent Leica 14–150mm zoom lens which is equivalent to a 28–300 zoom for 35mm. The difference is the Leica lens weighs 520 grams as compared to 1670 grams for the Canon equivalent and to my eyes the Leica lens is sharper. **Photographic Details: Olympus EM1, ISO 200, Leica 14–150mm f3.5–5.6 lens at 14mm (28mm on 35mm), 1/640 at f6.3.**

Above: Busy freight scenes have become increasingly rare across the UK, so I was delighted to capture this view at Didcot on 14 September 2015. 70002 winds through the station rather than using the avoiding lines, with a Southampton to Trafford Park Freightliner container train. 66160 shunts wagons for what will become the Didcot to Mossend wagonload freight, whilst two further Class 66s are stabled adjacent to the Class 165 that has stopped in the station. **Photographic Details: Olympus EM1, ISO 200, Olympus 40–150mm f2.8 lens at 40mm (80mm on 35mm), 1/640 at f4.5.**

Right: On 23 December, 2015 there was unbroken sunshine at Barnetby and quite a crowd had gathered to record the last 48 hours of the semaphore signals there. We had official visits to both Barnetby and Wrawby Junction boxes, but I also took some images of trains passing and of the crowds. A couple of dozen photographers jostle for a spot as 60065 passes with a Kingsbury to Lindsey empty oil train. Later in the day more than a dozen photographers with poles had to be removed from the waste ground opposite Wrawby Junction by the British Transport police and by sunset there were at least 50 photographers crowding the station platform for one last shot of the scene. **Photographic Details: Canon EOS7D, ISO 400, Canon 70–300mm f4–5.6 lens at 170mm (272mm on 35mm), 1/250 at f7.1.**

Right: On 8 January 2015, John Illingworth fills in the train register book at Barnetby East during one of our official visits to the area in the final year that it had mechanical signalling. John had joined the railway later in life, as a General Purpose Relief Signaller at Lowdham in Nottinghamshire. He then moved to a relief job on South Humberside and when the boxes there closed, became a relief signaller on Merseyside. Most boxes there closed at the end of 2018 when John became a relief signaller in the Crewe area. His goal of learning over 50 boxes before his retirement looks like being achieved! **Photographic Details: Canon 5D Mk III, ISO 800, Canon 24–105mm f4L lens at 24mm, 1/60 at f4.**

Right: On 28 September 2015, Trevor Maxted and I were granted an official tour of all the manual signal boxes in the Worcester area, which included a visit to Droitwich. The box here was built in 1907 and contains 79 levers, plus two for detonators. Resident signaller Alan is explaining aspects of the box's operation to Trevor Maxted, who himself now has over 50 years' experience as a signalman and is still working in 2021 as he approaches the age of 70. The better depth of field achieved by the smaller sensor is very evident in this view. **Photographic Details: Olympus EM1, ISO 200, Olympus 12–40mm f2.8 lens at 12mm (24mm on 35mm), 1/50 at f5.6.**

Above: During the March 2015 Telerail trip to Switzerland, I made the first of three visits to the Gotthard Pass, before the base tunnel opened in 2016, with full service commencing at the end of that year. On the south side of the pass an intermodal train is leaving the Pardorea Spiral at Faido. Phil Metcalfe and I found a spot in somebody's front garden for this view which shows a second intermodal train ascending the Pass on the second level of the spiral and the third level clearly visible above that. **Photographic Details: Olympus EM1, ISO 200, Olympus 40–150mm f2.8 lens at 40mm (80mm on 35mm), 1/640 at f5.6.**

Top right: One of the crazier trips in my 50 years was a weekend visit to China! Fellow photographer and BA pilot Steve Arthur had often visited China, but never discovered the last remaining steam, which was a three-hour train ride away in Fuxin. Everything came together one weekend as he was flying to Beijing and traded the return leg with a colleague. The same weekend another fellow photographer, John Birkinshaw, was working in China and had access to a car in Fuxin. A plan was hatched for two days in Fuxin, but we had to find something to do on the day of our arrival, so I suggested we drive to Shidu, three hours south-west of Beijing, as I had seen a picture of this amazing railway setting 15 years earlier and wondered if it might be worth a visit. Sadly there was an all pervading winter smog, even out in the countryside and the green diesels tended to blend in with their background. Nonetheless, we managed this one almost clear shot of an empty coal train passing the tunnels at Shidu on 25 January 2015. **Photographic Details: Canon EOS 5D Mk III, ISO 400, Canon 24–105mm f4L lens at 24mm, 1/640 at f7.1.**

Right: On 20 March 2015, Bern–Lötschberg–Simplon Railway locomotives 186 & 191 traverse the bottom level of the spiral at Wassen, Switzerland, which is 50 km away from the Pardorea Spiral, with a steel train from Emmenbrucke to Lecco. The three level loops at Wassen were one of my favourite locations on the Gotthard Pass with the church at Wassen standing proud on the hillside in the background. The second and third levels of the incline can be seen in the upper left corner of this view. **Photographic Details: Olympus EM1, ISO 200, Olympus 12–40mm f2.8 lens at 20mm (40mm on 35mm), 1/1250 at f4.5.**

Above: On 18 March 2015, a Tirano to St. Moritz train descends the Bernina Pass on the Rhaetishe Bahn. With a summit of 2328 metres, this is the highest Alpine railway pass. Typical of several workings along the line, passenger and freight traffic are combined on this train, with empty timber wagons tagged onto the rear of the consist. **Photographic Details: Olympus EM1, ISO 100, Olympus 12–40mm lens at 21mm (42mm on 35mm), 1/2000 at f5.6.**

2016

On the work front, the significant event of the year was working for six months as a consultant surgeon in Stockport, which led to a weekly commute by plane from Norwich to Manchester, leaving at 06.45 on a Monday morning and arriving in work at 08.30. Because of the hand baggage restrictions on the small propeller plane, I purchased a Lumix TZ-100 pocket camera to fit in my briefcase, along with clothes for the week. Fortunately, as a surgeon, much of my time was spent in theatre blues, so there wasn't much need for normal clothing! On the photographic front, images of 16 new signal boxes were added to my collection. I also made the last of my 33 visits to China to chase steam, with a short trip to Sandaoling with John Birkinshaw.

Above: I have included this shot as I think the site shown competes for the title of the most changed location during my 50 years of chasing trains. The line is the route from the former Radyr Quarry Junction to Radyr and the road overbridge is roughly where there used to be a bi-directional semaphore signal in the middle of the Up and Down Yards at Radyr. 66551 heads an empty coal train from Aberthaw to Tower Colliery on 22 June 2016. The main yard was to the left of the locomotive and the amount of change that this location has seen is evident when comparing this to the 1990 view of Radyr Yard on page 76, in which the signal mentioned above is visible in the upper left extremity. **Photographic Details: Olympus EM5 Mk II, ISO 200, Olympus 12–40mm f2.8 lens at 40mm (80mm on 35mm), 1/400 at f5.**

Above: The hospital at Stepping Hill in Stockport is the nearest acute hospital to Peak Forest and so was a great place to work during the summer when the long evenings allowed visits to Great Rocks signal box for coffee. On 22 July 2016, I had a free morning and visited nearby Hindlow to record the unusual sight of two trains in action. 66610 reverses a Tunstead to Briggs service into the sidings at Briggs, whilst 66058 waits to depart with a Dowlow to Ashburys loaded stone train. While standing on the bridge, I was accosted by a security guard who said we he was concerned for my safety, because of the passing road traffic. Fortunately, we had a nice conversation and he allowed me to stay long enough to record this scene. Health and Safety considerations in the UK, as well as misplaced security concerns, have become the bane of railway photographer's lives, making taking railway pictures abroad much more attractive. **Photographic Details: Panasonic Lumix TZ100 at 48mm (130mm on 35mm), ISO 200, 1/200 at f7.1.**

Top right: The one-inch sensor of the Panasonic Lumix TZ100 is a relatively recent innovation, giving a 2.7 times shorter focal length compared to a full frame sensor, which is the size of a 35mm negative. On 23 March 2016, during my first week with the camera, I popped to March to put it through its paces. 66721 is seen passing March East with the Middleton Towers to Goole sand train. The quality at ISO 800 convinced me there was a place for a one-inch sensor pocket camera in my briefcase! **Photographic Details: Panasonic Lumix TZ100 at 24mm (66mm on 35mm), ISO 800, 1/640 at f5.**

Left: The autumn colours add to this shot, which was taken on 18 October 2016 from a fairly rare location, Greenford East signal box. I had arranged an official visit with Trevor Maxted, as this was the last mechanical signal box in the South East that I needed to visit for my Signal Box Encyclopaedia which was published in 2020. The signaller on duty had only qualified a few weeks earlier and there was a track circuit fault during our visit which Trevor tried to help with. We were very fortunate in that one of the few freights of the day came past when we were there, in the shape of 66750 on 6Z58, the Calvert to Cricklewood empty spoil working. The box still stands, with its 76 levers and is largely hidden from view behind trees and palisade fencing. **Photographic Details: Olympus EM1, ISO 200, Olympus 40–150mm lens at 120mm (240mm on 35mm), 1/500 at f5.6.**

Above: DB Cargo Class 185, 185 103 and a sister locomotive climb onto the third level of the spiral at Wassen with a Wuppertal to Piacenza intermodal service. The train is banked by a third Class 185, 185 096. The second level can be seen on the right of this image and the smaller photograph shows the crowd on the hill capturing the last weeks of the Gotthard Pass on 7 May 2016, as the same train is propelled up the second level by 185 096. **Photographic Details: Olympus EM1, ISO 200, Olympus 12–40mm f2.8 lens at 12mm (24mm on 35mm), 1/800 at f5.**

Right: After working in Hull during 2012, I had a long association with a number of railway employees in the area, whom I continued to visit several times each year. This view from the footbridge at Gilberdyke station is unusual in that there are four mechanical signal boxes to be seen in the form of Oxmardyke, Broomfleet, Cave Crossing and finally Crabley Creek. The faint impression of the bridge across the tracks several miles away is the road bridge to the west of Brough station. 158861 whisks up the autumn leaves as 66091 passes Broomfleet station in the distance with the Hedon Road to Rotherham steel train. The ability to use ultra telephoto lenses on the micro four thirds format was another attraction to changing away from 35mm. **Photographic Details: Olympus EM-1, ISO 800, Leica 100–400mm f4-6.3 lens at 400mm (800mm on 35mm), 1/160 at f6.3.**

Left: I am sure many photographers have spotted a location which has worn away at them, because it looks like it should provide a spectacular image, but seems impossible to access. The Trisanna Brucke, to the west of Landeck in Austria was like that for me. I had driven under it many times on my way to ski in Galtur and Ischgl, but never been able to see how to get a picture. On 20 September 2016, after a walking holiday in Galtur, I decided to try and access the embankment to the west of the bridge. With no way to climb the steep heavily wooded hillside, I had to use the walkway beside the track on the bridge to get there. I was rewarded with this view of an OBB Railjet service from Vienna to Zurich headed by Class 1116 1116 209. **Photographic Details: Olympus EM1, ISO 200, Olympus 40–150mm f2.8 lens at 60mm (120mm on 35mm), 1/640 at f5.6.**

Left: The weather is very rarely this clear at Sandaoling in the far west of China. This picture is the result of the kindness of Arron Homewood, a fellow photographer. It was my first trip to China using the Olympus and I found to my horror that below minus 10 degrees Celsius the electronic shutter was sticking. This amused all the Canon and Nikon users on the trip immensely. Arron kindly lent me a spare Canon 7D body with a standard lens for the start of each day which allowed me to capture this JS Class locomotive hauling coal out of the pit at Sandaoling with the "Heavenly Mountains" in the background. It was even more amusing that David Hill and his fellow travellers had not bothered to come to the pit on 18 January 2016, as they were taking pictures of "Mrs Flower Pot" their Uighur guide who was ice skating and by the time they joined us the weather had closed in. I later found out that the shutter on my new Olympus was faulty and with a new mechanism it worked fine at low temperatures. **Photographic Details: Canon 7D, ISO 200, Canon 50mm f1.4 lens, 1/500 at f7.1.**

Below: In February 2016, Steve LeCheminant and I set up a small exhibition at the Brief Encounter café in Carnforth to commemorate Vernon Murphy's marvellous 1960s steam photography. It was great that his wife and one of his daughters were able to attend and see his photos displayed.

Left: Trains were running to a local coal stocking site during our visit in 2016 and on 15 January 2016, a JS Class locomotive is captured hauling coal for local consumption just as the sun sets. It is rumoured that the mine and/or railway line to Sandaoling will close, so it looks as if these images will be my last of steam in China, which I visited 33 times over 25 years. **Photographic Details: Olympus EM1, ISO 200, Olympus 12–40mm f2.8 lens at 32mm (64mm on 35mm), 1/400 at f5.6.**

2017

Life as a consultant surgeon was winding down in 2017, as the waiting list work in Taunton dried up following a change in government policy whereby A&E waiting times seemed to take priority over waiting time for surgery. On a personal front, a retirement hobby of winemaking began to become accidentally successful and so I decided to retire from the NHS and become a winemaker! I was able to visit the Sandstone Railway in South Africa and join a superb Farrail trip to Burma, which was based around the British semaphore signals at Bago. A further slightly crazy trip was to the last three surviving hump yards in India. After detailed research using The Great Indian Railway Atlas and Google Earth, it looked like three hump yards were still in operation. I was able to get permission from Indian Railways to visit these, only to find that just two of them were still working. The resulting free time was spent visiting the city of Varanasi on the Ganges. In the UK, I visited 65 new signal boxes for my work on the Signal Box Encyclopaedia. This included a trip to Scotland to fill in some gaps and while there I revisited Wormit, where I had first taken pictures in 1990 when I was learning laparoscopic surgery at Dundee Ninewells Hospital.

Below: This classic view of the Tay Bridge seemed to have opened up a bit in 2017, because a derelict piece of land could be accessed to gain a little extra height. I am not sure whether it was part of someone's garden or a potential building plot, but nobody seemed to mind me standing there for a while to capture 60002 heading south with the Aberdeen to Oxwellmains empty cement working on 11 May 2017. **Photographic Details: Olympus EM1 Mk II, ISO 200, Olympus 40–150mm lens at 150mm (300mm on 35mm), 1/500 at f6.3.**

Right: This new signal at Oulton Broad North had a very short life, yet it cost Network Rail a fortune to install. In order for trains to arrive on the Norwich-bound platform and keep the crossing barriers open on the busy road to the west of the station, signal OB30 had to be moved 50 yards closer to Lowestoft. A second consequence of the move was the need to extend the platform at the Lowestoft end of the station. In all the bill came to nearly £1 million – then the Wherry lines were resignalled in 2019, rendering the semaphore redundant! 68003 & 68016 "top and tail" the 2D70 Norwich to Lowestoft working on 10 March 2017. **Photographic Details: Olympus EM1 Mk II, ISO 400, Olympus 40–150mm lens at 85mm (170mm on 35mm), 1/320 at f11.**

Left: During 2017 I had regular breakfast "meetings" with relief signalman Owen Bushell. Given his vegan habits, I had to pick vegetables from our garden, otherwise he would allow eggs from our free-range chickens. We had breakfast in Brundall Junction, Oulton Broad North, Yarmouth and my favourite signal box seen here, Reedham Junction. Much of my railway photography is about people and places and Reedham Junction is where I first met Steve Ashling in 1996. Steve then introduced me to Owen in 2011. Owen was the last signalman appointed to a box between Norwich and Ely when he joined as a resident signaller at Eccles Road. He then became a relief signalman on the Wherry lines until they were resignalled, at which point Owen left the railway and went to live in Austria! He seems happy with his breakfast on 9 September 2017. **Photographic Details: Olympus EM1 Mk II, ISO 400, Olympus 12–100mm f4 lens at 12mm (24mm on 35mm), 1/13 at f5.**

Above: My exploration of operational hump marshalling yards around the world led to a brief visit to India in February 2017. The large yard at Mughalsarai was touted as the largest in Asia, before the rise of China, which now boasts two dozen hump yards, including the largest in Asia at Zhenzhou North. Meanwhile, wagonload traffic in India has been in decline, leaving just two active hump yards. Mughalsarai has been drastically rationalised from two hump yards to one, which is now underutilised. The hump yard at Andal Junction is probably a bit busier and offers the fascinating sight of a hump controlled by British heritage manual signalling. The signal box controlling the hump approach is seen here on 5 February 2017. Beyond the track diagram is a panel indicating which sorting siding has been selected by the signalman in the hump tower which is a couple of hundred yards away. **Photographic Details: Olympus EM1 Mk II, ISO 200, Olympus 12–100mm f4 lens at 12mm (24mm on 35mm), 1/15 at f4 (HDR).**

Top left: During an excellent trip to Burma that was organised by Farrail, Bernd Seiler even hired passengers for our train! On 1 November 2017, YD Class 2-8-2 No. 967 is seen crossing the river at Kyaikhto with a recreation of the Mokpalin to Mottama mixed train which used to run in the 1980s. The fake passengers stand conveniently in the carriage doorways whilst the young lad deserves a BAFTA for his acting, as he passes the anchor to his father. **Photographic Details: Olympus EM1 Mk II, ISO 200, Olympus 12–40mm f2.8 lens at 28mm (56mm on 35mm), 1/640 at f6.3.**

Left: On 2 November 2017, No. 967 is captured passing the Pagoda at Yinnyein with our Thaton to Mokpalin train. I included this image both because the trip to Burma included multiple attempts to get the perfect "pagoda plus train shot", but also because it is two images merged together. I do not often manipulate images using Photoshop, but the fisherman in the main boat passed in front of me just 30 seconds before the train passed over the bridge and I have been able to take the bottom half of the earlier image and put it with the top half of the one with the train passing. I think it's quite hard to see the join! **Photographic Details: Olympus EM1 Mk II, ISO 200, Olympus 12–40mm f2.8 lens at 13mm (26mm on 35mm), 1/640 at f5.6.**

2018

The big photographic development of 2018 was the purchase of a drone. I bought a DJI Mavic Pro with its 12 megapixel sensor (very similar to the sensor in a Canon G15). What a revelation. Drones have opened up so many new possibilities and pictures taken with them have made up a high percentage of my output from 2018 onwards. Writing this just days after I hear that 60009 "UNION OF SOUTH AFRICA" has had a major breakdown and that's its owner John Cameron is in hospital after a bad accident, reminds me of a great day out in 2018 when I was allowed to ride on the footplate of "09" from Newcastle to Berwick-upon-Tweed. On the signal box front, 2018 saw the pace of box closures pick up, with 19 mechanical boxes reaching the end of their lives. It was also the last time that I spent a day out on the Wherry lines, before its boxes closed in 2019. On the international front, I joined Farrail for a superb tour in Sri Lanka, managing to fall and break two ribs on the third day of the trip, but thanks to some interesting East German pain killers, I only missed a couple of hours of photography. Later research revealed that the tablets were a strong morphine derivative which is banned in most of the EU, but because it was invented by a German chemist in the 1920s, it can still be purchased in Germany!

Above: The Olympus 7–14mm (14–28mm on 35mm) is an ideal lens for cramped locations like the footplate of "Union of South Africa". On the right, John Cameron is in his favourite spot checking our schedule as we sit in a loop on the Northumberland coast, to allow a Kings Cross to Edinburgh express to overtake us. Back in the comfort of the dining car, Tim Littler and Wick Moorman have saved some bottles of wine and whiskey for the journey back from Edinburgh to York! **Photographic Details: Olympus EM1 Mk II, ISO 200, Olympus 7–14mm f2.8 lens at 7mm (14mm on 35mm), 1/13 at f3.2.**

Above: I made my last visit to the Wherry lines on 18 October 2018, which was a glorious autumn day. The final train I photographed that day was the railhead treatment train coming off the Berney Arms line at Reedham Junction. With St. John the Baptist church visible on the horizon, 37423 & 37606 join the main line from Lowestoft to Norwich. **Photographic Details: Olympus EM1 Mk II, ISO 200, Olympus 12–40mm f2.8 lens at 40mm (80mm on 35mm), 1/800 at f5.**

Left: The railhead treatment season brings a much-welcome variety of motive power to the network and for many years the diagram covering north and south Humberside used Class 20s, due to weight restrictions along parts of the route. Having seen increasing numbers of excellent images taken by photographers pushing the ISO rating of their cameras to the limit, I decided to try some night photography of a moving train, something I have hardly ever done. To do this I also invested in an f1.2 standard lens to hopefully get a quick enough shutter speed to "freeze" the train. I chose the wonderful train shed at Beverley on a freezing night in October 2018. 20303 & 20305 trundle through the station on their way to Bridlington where they will reverse and return to York. **Photographic Details: Olympus EM1 Mk II, ISO 3200, Olympus 25mm f1.2 lens, 1/200 at f1.2.**

Above: Having practised with my drone in the back garden, my first outing with it was on 27 June 2018. The open fields and countryside at Reedham Junction seemed a good place to try and capture my first railway image. 37407 & 37425 cross the swing bridge hauling the 14.55 Norwich–Lowestoft service. Flying over the water was a bit nerve racking as when starting out with a drone there is always the fear it may crash or fall out of the sky, however this is usually quite unfounded and so far, my drones have always returned home safely! **Photographic Details: Mavic Pro, ISO 100, 4.7mm (26mm on 35mm) lens, 1/1500 at f2.2 one third stop under exposed.**

Right & top right: On 2 February 2018, we arrived at Diyatalawa station in Sri Lanka on the "railway to the moon", so called because the summit between Nanu Oya and Badawalla is 6226 feet in elevation. There was thick mist and things didn't look promising. Climbing towards the summit the mist cleared slightly, allowing this shot of our train bursting into the morning sun. B2B Class 4-6-0 No. 213, which was built at the Vulcan Foundry in Newton-le-Willows and exported in 1925, made several runpasts at this spot, but the first with the mist at its thickest was the image that captured the scene most dramatically. The muddy water channel in the foreground was packed with leeches as several of my fellow photographers found to their cost. From left to right, the smaller photograph shows John Birkinshaw, Pietro, myself and John Parker. I look slightly ill at ease, as this was taken just hours after I broke my ribs when falling downhill. **Photographic Details: Olympus EM1 Mk II, ISO 200, Olympus 12–40mm f2.8 lens at 19mm (38mm on 35mm), 1/1600 at f9.**

2019

Having moved to Suffolk in 2018, I retired fully and had more time for photography. Most of my UK trips were made with Steve Arthur and Martyn Fordham, both avid contributors to the photograph sharing website Flickr. One particularly excellent two-day outing took in Worcester, Shrewsbury and Crewe. I also revisited South Wales and was taken aback by the many changes that I found had been made to the railways of the Principality which I knew so well in the 1970s and 1980s. When it came to international travel, I am reminded of some words that my dear friend Vernon Murphy uttered just weeks before he died. “We've had the best time on the planet, with affordable air travel, steam engines and decent cameras, it's all downhill from here.” Sad though his words were, there was some truth in them. I joined a superb Farrail trip to South Africa to photograph “The Red Devil”. The country is now almost impossible to visit and the pandemic led to the collapse of plans for further steam charters there, not to mention huge financial losses for the tour operators. I enjoyed a crazy four-day visit to India to record the last large collection of semaphore signals on the Indian railways at Dhurri Junction. These were replaced just weeks after our visit. Finally, I had an amazing few days photographing the “Big Boy” in the USA, on its first outing after it had been returned to operational condition. This was a scene that many never dreamt would happen and thanks to the kindness of Wick Moorman, I had a VIP visit to the footplate and a lengthy chat with Ed Dickens who headed the project to bring this iconic steam giant back to life. Little did we know what was round the corner, as the forthcoming global pandemic would mean that it would be another two years before it was possible to travel to the USA again. Finally, in October 2019, on a visit to the Hannover area with Steve Arthur and Guy Houston, our hire car was broken into at nearby Lehrte and all my camera gear was stolen, as was much of Steve Arthur's equipment. Vernon's words echoed in my mind at that point, although worse was to come with the subsequent lockdowns and a freeze on virtually all international travel.

Below: Trackside vegetation has increased steadily over the last 50 years since the end of steam. When the widespread use of weedkillers stopped in the 1990s, this allowed an almost relentless march of vegetation. Gone are the days when Class 20s appeared on the weed killing train; now heritage motive power is confined to the much more environmentally-friendly railhead treatment trains. One plant that has thrived above all others is Buddleia. When revisiting Llantrisant on 17 July 2019 (the station there is now called Pontyclun), it was even growing out from the bridge parapet, so I decided to make a feature of it, rather than remove it. 70807 waits in the loop with the empty log train from Chirk to Baglan Bay. **Photographic Details: Olympus EM1 Mk II, ISO 200, Olympus 12–100mm f4 lens at 50mm (100mm on 35mm), 1/30 at f20.**

Above: On 18 September 2019, we had just enjoyed a superb chips with curry sauce at the chippy opposite Crewe's Gresty Road football ground and decided to fly our drones over the waste ground to the south. An unidentified Freightliner Class 66 winds north towards Salop Goods Junction with a Basford Hall to Trafford Park intermodal train. The outline of the once mighty hump yard at Basford Hall can be seen behind the train, whilst electric multiple units withdrawn from the King's Lynn line fill the sidings in the centre of the image. **Photographic Details: Mavic2 Pro, ISO 100 two thirds under exposed, 10.3mm (28mm on 35mm) lens, 1/500 at f5.6.**

Below: A question which will no doubt exercise the minds of young railway enthusiasts is "Can I cop a locomotive if I only saw it with my drone?!" Back in the 1970s it was a simple life, you either saw a locomotive's number or you didn't. Views like this highlight a new dilemma; a range of Direct Rail Services' motive power is seen on Gresty Lane depot on 18 September 2019, invisible from the roadside, but clear as day on my smart controller screen. **Photographic Details: Mavic 2 Pro, ISO 100 two thirds under exposed, 10.3mm (28mm on 35mm) lens, 1/500 at f6.3.**

SALOP GOODS JUNCTION
88008
0845 266 6535

Above: The beauty of the Welsh valleys and the scars of coal mining are evident in this view of Onllwyn Washery, taken on 1 May 2019. I was delighted to meet up with Chris Davis, after we made contact through Flickr and spend a couple of days exploring old haunts. Perhaps the biggest shock was that only two coal services ran in the whole of South Wales during my visit. This working, which travelled from Margam to Onllwyn and then on to Immingham Sorting Sidings, carrying coal for export, was hauled by 66023. The second working, from Cwm Bargoed, carried metallurgical coal to Port Talbot and attracted the attention of environmentalists. How it is better to import metallurgical coal from the other side of the planet, rather than let it free wheel down the valley to Port Talbot is beyond me. The suggestion that we don't need steel and can start making cars and bridges with bamboo is equally ridiculous. So for now, Britain still has two steel works and a tiny coal production industry. **Photographic Details: DJI Mavic 2 Pro, ISO 100, 10.3mm f2.8 (28mm on 35mm) lens, 1/400 at f5.6.**

Left: To me this image is a sign of the future. The main picture is taken using an iphone 8 plus, whilst the two inserts were taken with an Olympus camera. John Illingworth is on the phone, as a colleague "learns the box" at Salop Goods Junction. One insert shows 88008 passing, with a Mossend to Daventry container service, whilst the other shows the marvellous wooden block instruments which are still in use today. We had permission to visit the box from the Local Operations Manager and had a marvellous few hours as freight after freight passed us on the Crewe avoiding lines. For a point and shoot camera, the iphone has done a remarkably good job and cameras on iphones are only going to keep getting better. **Photographic Details: iphone 8 plus, ISO 40, 4mm (28mm on 35mm), 1/17 at f1.8.**

W68
W69
W70
W71
72W
WDG4G
साबरमती
SABARMATI
WDG4
12909
1

Above: Farrail Tours arranged a magnificent South Africa trip in 2019, centred around the return to steam of the South African Railways Class 26 4-8-4 which is known as the Red Devil. The remarkable locomotive, modernised by the famous steam engineer L. D. Porta, is seen here on its way back from Ashton to Worcester on 20 June 2019. **Photographic Details: Olympus EM1 Mk II, ISO 200, Olympus 12–100mm f4 lens at 44mm (88mm on 35mm), 1/400 at f5.**

Left top & bottom: I think my most remarkable discovery of 2019 was the surviving signals at Dhurri Junction. Again, the benefits of the internet, especially Flickr, were pivotal in my trip there. I had been following Don Gatehouse on Flickr and he posted images of the remarkable array of signals at Dhurri. Don had contributed to our Freight Only books back in the 1980s and so I messaged him through Flickr to ask about the location. As a result, Steve Arthur and I made a short trip to photograph the signals in March 2019, just weeks before they were demolished. Two signal boxes with 80 and 60 levers controlled the station area. The monochrome image shows Dhurri West, with the point rodding leading from the signal box as dawn breaks. Later that day, a westbound container train stops at Dhurri station to allow the brand new WDG4G diesel locomotive to be attached to the front of the train before it heads to Bathinda Junction. Access to the signal boxes and station would not have been possible without Tim Littler and his contacts in Earthen Experiences. **Photographic Details: First (mono) image: Olympus EM1 Mk II, ISO 200, Olympus 12–100mm lens at 20mm (40mm on 35mm), 1/800 at f6.3.**
Second (colour) image: Olympus EM1 Mk II, ISO 200, Leica 100–400mm f4–6.3 lens at 318mm (637mm on 35mm), 1/160 at f9.

Above: After its Thai owners withdrew their rescue bid, Redcar steelworks had stood silent for a couple of years and in 2019 it was rumoured that demolition of the plant was imminent. The view across the Coatham Marsh Nature Reserve is one I had not previously taken, even though I had lived in the North East for six years. It was also a view I wanted to capture in order to make a back scene for my model railway, hence my visit there on 8 August 2019. 66017 heads the lunchtime trip from Tees Yard to Skinningrove, a freight that I had first photographed back in 1985. **Photographic Details: Olympus EM1 Mk II, ISO 200, Olympus 12–100mm f4 lens at 66mm (132mm on 35mm), 1/800 at f5.**

Right: Perhaps the most impressive sight I have ever encountered in 50 years of chasing trains was the Big Boy at full chat, climbing through Echo Canyon with the breakfast special from Ogden to Evanston on 12 May 2019. 25 of these powerful articulated 4-8-8-4 steam locomotives were built by the American Locomotive Company in the 1940s, but their working lives were relatively short, as they had all been retired by 1962. One of the eight survivors, no. 4014, was returned to operational condition in 2019, making it the world's largest and most powerful operational steam locomotive. There was to have been a highly organised trip, involving friends from the USA and a private railcar from Denver to San Francisco, but unfortunately that was cancelled. I was therefore very grateful to be able to join John Parker and Ben Kletzer on a last-minute five-day visit to the USA. We had tickets to dine on this train but that would have meant not taking any pictures of it on this epic journey. Ed Dickens, neckerchief perfectly in place, leans out to wave to us at the appointed spot. **Photographic Details: Olympus EM1 Mk II, ISO 200, Olympus 12–100mm f4 lens at 17mm (34mm on 35mm), 1/800 at f5.6.**

Next pair of pages: My thanks for talking me into photographing at this spot and advising on it go to Chris Davis. Porthkerry Viaduct was a location where I took many pictures in the 1970s, as we often had church picnics in the park which lies in its shadow. In those days the line was busy with MGR trains to Aberthaw Power Station (now closed) and there was also a car parts train which ran every evening from the factory at Bridgend to Dagenham. On 16 July 2019, 66093 heads west with a coal train from Cwm Bargoed to Port Talbot Steelworks. This is only the third image in the book which is a cheat – the cows in the foreground decided to all move a few minutes before the train passed, so I merged the top and bottom sections of two photographs to create this view. **Photographic Details: Olympus EM1 Mk II, ISO 200, Olympus 12–100mm f4 lens at 61mm (122mm on 35mm), 1/500 at f7.1.**

EWS

EWS

The 2020s

2020

What a year. I would never have guessed when my wife came home from skiing at the start of March 2020, with what seemed to be a bad cold, that she had a new disease called COVID-19. We were due to return to Austria ten days later, but the first lockdown put an end to that. Whatever the rights and wrongs or people's views are, so far the 2020s have been a big challenge. Most of our family had COVID, several of whom tested positive with no symptoms. I failed to get it in 2020, in spite of being coughed on for weeks on end by people who tested positive for the virus. I then came out of retirement to help with vaccinations and opportunities to travel and take photographs were severely hampered.

During 2020, each time the lockdown restrictions were lifted a little, Steve Arthur, Martyn Fordham and I made trips out, the best of which was to Northumberland to photograph the biomass trains to Lynemouth Power Station. Most places we visited were ghostly quiet, like Ipswich station in June, when there were more staff than passengers. Between July and September we made three trips to Germany to record freight traffic for our "Freight Only Germany" series. But as autumn came, a second lockdown was imposed and 2020 ended with the only trains to photograph being on my model railway!

Above: There was a bit of a party atmosphere on the bridge at Kirton Lime Sidings on 23 June 2020. A dozen or so photographers were there to record 66131 on a Drax to Immingham biomass train, diverted via the Brigg route. Fish and chips in hand from the local chippy, this shot was taken one handed to keep grease off the camera! **Photographic Details: Sony AR7 Mk IV, ISO 200, Sony 24–105mm f4 lens at 30mm, 1/800 at f5.6.**

Left: This view illustrates Britain's industrial decline. The horizon shows the closed and now-demolished Ferrybridge Power Station, as well as the closed Eggborough Power Station. Heading across the picture is 66004, with the daily train from Scunthorpe which takes steel blooms for processing in Lackenby. This is only necessary now that steel production has ceased in Redcar, leaving the UK, the world's fifth biggest economy, with just two operational steelworks. In the foreground is 66012 returning to Immingham with an empty biomass train from Drax, on its way to pick up more pelletised Scandinavian trees to burn in the furnaces at Drax Power Station. So stable is the DJI drone that this is a merger of two images taken 90 seconds apart – can anybody see the join? **Photographic Details: Mavic 2 Pro, ISO 100, 10.3mm f2.8 (28mm on 35mm) lens, 1/5000 at f4, one stop under exposed.**

Above: Once a week there is a short freight train along our local line through Thurston, the Tuesdays only Ipswich to Lindsey empty oil tanks, which returns fuel oil tanks from the depot at Ipswich. On 16 June 2020, 66505 passes fields of winter wheat at Haughley Junction. **Photographic Details: Sony AR7 Mk IV, ISO 200, Sony 24–105 f4 lens at 24mm, 1/800 at f8.**

Top right: Ipswich station on 16 June 2020 had an apocalyptic feel to it; there were no passengers and dozens of staff walking round aimlessly. The car park, from which this shot was taken was completely empty. The staff were friendly and allowed us onto the station to photograph trains, all of which were empty. The sight of London to Norwich express trains without a single passenger was quite bizarre. 66548 is seen passing through the station with a Felixstowe North to Lawley Street container service. **Photographic Details: Sony AR7 Mk IV, ISO 100, Sony 24–105 f4 lens at 46mm, 1/500 at f5.6.**

Right: On 7 November 2020, the level of restrictions in place permitted going out for a walk and so we went for a walk along the Hundred Foot Bank in Cambridgeshire. 66792 is captured on 6E88, a sand train from Middleton Towers to Goole. On the extreme right of the image another drone can be spotted. The "RAF" were out flying that day (Rhodes, Arthur, Fordham)! **Photographic Details: Mavic 2 Pro, ISO 100, 10.3mm f2.8 (28mm on 35mm) lens, 1/5000 at f4 (one stop under exposed).**

Freightliner
66548
40

Above: Taking pictures with drones has been relatively trouble free for the last three years. Having registered the larger Mavic 2 Pro drone and completed the relevant online training, we had not anticipated any problems when we visited Godnow Bridge on 14 May 2020. The aim had been to capture the tank train alongside the Stainforth and Keadby Canal and although the sun dipped behind a cloud at the last minute, "the RAF" were all pleased with their pictures. Suddenly a large and very irate lady stormed up to us and accused us of flying over her property, which we weren't. As suggested by the drone code, I offered to show her the image I had just taken to which she leapt back, for fear of catching COVID! She then said she was "getting some blokes to come and do us over" at which point we realised there was no way to

reason with her and we beat a hasty retreat to another location. Passing her house, which is out of view in this picture of 70806 working the tank train to Preston, we glimpsed her sunbathing in her garden, semi-naked – no wonder she was so wound up! **Photographic Details: Mavic 2 Pro, ISO 200, 10.3mm f2.8 (28mm on 35mm) lens, 1/2000 at f4 (one stop under exposed).**

Inset: This is what railway photography looked like during 2020!

Above: Of the trips to Germany in July, August and September 2020, the most enjoyable was in August when Guy Houston, Steve Arthur and I spent three days staying in Koblenz and taking pictures of freight trains along the river Rhine. My notebook records that we photographed 111 freights in three days! On 12 August 2020, Steve and I had decided to remain on the left bank of the Rhine at Kaub, with the plan to try and take pictures at Oberwesel and St. Goar. Guy in contrast had noted that there were three times as many freights scheduled along the river's right bank so took the ferry to Kaub and planned to travel by train to Koblenz for supper that evening. Guy is

visible in the bottom right corner of this image, at the end of the station platform, busy trying to get skin cancer, but also recording all the locomotive numbers on the passing freights. Steve and I were on the other side of the river and used our drones to capture the majestic Castle at Kaub with the trains in view. Here we see Class 185s 185092 & 185245 with EZ51467, a wagonload freight from Gremberg to Kornwestheim. **Photographic Details: Mavic 2 Pro, ISO 100, 10.3mm f2.8 (28mm on 35mm) lens, 1/1600 at f3.6 (two thirds stop under exposed).**

Above: With Greater Anglia retiring their fleet of Class 90 electric locomotives and Freightliner buying them, the days of the venerable Class 86 were numbered. The evening Felixstowe North to Trafford Park freightliner container service can only be photographed as it climbs Belstead Bank, on the way south from Ipswich, for a couple of weeks either side of the summer solstice, so on a sunny 16 June 2020 it was good to capture 86609 & 86622 on this train. Noticeable during 2020 was the shift in traffic density, such that more container services from Felixstowe used the cross-country route through Ely, rather than the formerly busier route round the north London line. **Photographic Details: Sony AR7 Mk IV, ISO 200, Sony 70–350mm f4.5–6.3 lens at 156mm (234mm on 35mm), 1/640 at f5.6.**

Right: Having written two tomes on British marshalling yards and two books on American railyards, is there a book coming out on European marshalling yards you ask? The answer is that that is certainly my intention! Germany still has over 20 operational automated hump yards, many of which handle 2000–3000 wagons every day. One of my favourites is Nurnberg, seen here from the air. Class 185 185634 passes under the hump on the freight avoiding lines used by block workings and container trains. It is hauling DGS 43183, a freight from Zeilitz StW B2 to Targu Mures in Romania, which is running 926 minutes late! The yard at Nurnberg was totally rebuilt and modernised in 1988 and now has 60 primary sorting sidings, down from a pre-rebuild size of 106. It is still one of the busiest wagonload yards in Germany, with a capacity to sort up to 4700 wagons each day. **Photographic Details: Mavic 2 Pro, ISO 100, 10.3mm f2.8 (28mm on 35mm) lens, 1/1600 at f3.5 (two third stop under exposed).**

Above: I first photographed Wansbeck Viaduct, which crosses its namesake river to the south of Ashington, back in 1980, so it was great to revisit the location on 21 July 2020 with a drone. On the far left of this view is Marcheys House signal box. 60026 crosses the bridge with a Tyne Dock to Lynemouth Power Station biomass service. **Photographic Details: Mavic 2 Pro, ISO 100, 10.3mm f2.8 (28mm on 35mm) lens, 1/1250 at f3.5.**

2021

My 50th year photographing railways has been strange. The second lockdown, which took up much of the first half of 2021, was if anything more disruptive than the first one, as half a dozen international trips planned for the end of 2020 and the first half of 2021 all had to be cancelled. During the year we did manage a trip to Germany in August, but the logistics of all the testing and form filling made regular international travel a bit of a struggle. At home there were some great days out as reflected in my image selection for the year, but the "new normal" was more about staying at home and tending the vines, than gallivanting round the globe to photograph trains. On the positive side, my wife and I had our third booster vaccination to combat the virus and I managed to visit Pakistan with Farrail Tours in December. As we began to learn to live with COVID-19, some semblance of balance and normality started to return. In the meantime at least, we have a railway line at the bottom of our garden here in Thurston. It really is a case of coming full circle, as I now have lots of time on my hands as a pensioner, but international trips, which were very difficult to make in my youth, are now difficult again, but for different reasons.

Below: On 21 April 2021, the British Steel Lackenby to Scunthorpe Entrance E carried loaded steel beams on the front of the train, which is quite unusual as it normally returns to Scunthorpe empty to pick up steel blooms for processing in Lackenby. These beams are from Skinningrove and for export via the daily Scunthorpe to Dollands Moor service. 60001 crosses Yarm Viaduct, a structure I have often photographed, but never with such a perspective, now rendered possible by drones. **Photographic Details: Mavic 2 Pro, ISO 100, 10.3mm f2.8 (28mm on 35mm) lens, 1/1250 at f5, (one stop underexposed).**

Right: This is the location of the photograph below and illustrates how intrusive pole photography can be. It is visible in the upper centre area of the next image.

Below: The 22 April 2021 was a beautiful spring day with wall-to-wall sunshine and as a result there was quite a crowd along the clifftop at Brotton, on the Saltburn to Boulby line. This was also the last week that Freightliner were in charge of the potash trains, as these reverted to DB Cargo a week later. 66532 rounds the curve at Brotton with 6F26, a Boulby to Tees Dock potash working. The coastal erosion is very evident here and it looks as though it will only be a matter of time before major engineering work will be needed. On the other hand, a new pipeline from Boulby to Teesside may remove the need for a railway line here, as steel to and from Skinningrove could conceivably use road haulage. **Photographic Details: Mavic 2 Pro, ISO 100, 10.3mm f2.8 (28mm on 35mm) lens, 1/640 at f5 (one stop underexposed).**

Right: Never work with children or animals! The idea was to get Nellie (our golden retriever) and Archie (our Cairn Terrier) to smile at the camera as this Lawley Street to Felixstowe Freightliner working came past Thurston. No chance! 66541 made so much noise that both dogs turned – Archie to fight it and Nellie to run from it. At least my wife Irene followed my artistic directions to "smile at the camera". **Photographic Details: iphone 12 Pro Max, ISO 32, 5.1mm (26mm on 35mm), 1/200 at f1.6.**

Below: The march of technology has transformed the world of photography, making this combination of images possible. On 29 July 2021, I stood by the level crossing north of Ely to photograph 66560 heading north on a Felixstowe to Leeds intermodal service. At the same time, I had positioned my drone over the nature reserve to the east of the main line and captured the same train with Ely cathedral in the background – two shots for the price of one.
Photographic Details:
Main image: Mavic 2 Pro, ISO 100, 10.3mm f2.8 (28mm on 35mm) lens, 1/1000 at f5.
Secondary image: Sony AR7 Mk 4, ISO 200, Sony 70–350mm f4.5–5.6 G OSS lens, 1/320 at f10.

Below: One aspect of photography in the UK which I have completely failed to mention so far is the wonderful resource of Realtime Trains. Without this website, pictures like this would never be possible. On 31 March 2021 I noticed that a Mountsorrel to Carlisle ballast train was scheduled to pass Alrewas – the only train during daylight that day. The location is always a difficult one, made more so by the electrical interference when flying a drone here, such that my drone kept looking at its own feet with "gimbal motor overload" during this flight, even though it was over an empty field and away from any structures, except for a radio mast which may have been the cause. By repositioning the drone to look out of the right corner of the field of view, I managed to get this picture of 70814 at what was a new freight only location for me. **Photographic Details: Mavic 2 Pro, ISO 100, 10.3mm (28mm on 35mm) lens, 1/1250 at f4.**

Right and below: On 29 July 2021 I found myself in much the same position as I had done in 1972, sitting with two friends next to a busy mainline, enjoying the craic and the sunshine, and waiting for a train to appear. 90009 & 90012, still in Greater Anglia colours, pass Antsy on the West Coast Main Line with 4M87, the Felixstowe to Trafford Park Freightliner service. Our refreshment was our Hawkswood 2020 Pinot Noir, whereas it was a cup of tea from the station buffet in 1972. Our cameras are now digital, we fly drones and we take selfies with the iphone 12, all of which was science fiction when I started this journey 50 years ago. In the selfie, from left to right are myself, Steve Arthur and Guy Houston. One coincidence from this is that the iphone selected ISO 25, the same film speed as the legendary Kodachrome 25 ASA that was so favoured by photographers back in 1972.
Photographic Details: iphone 12 max pro front camera, ISO 25, 2.7mm (23mm on 35mm), 1/700 at f2.2.

The next few images were taken in north-west Pakistan during a ten-day Farrail trip in December 2021. 35 photographers from the UK, Germany, Switzerland, Belgium, Italy and the USA met in Peshawar for six days of photography on the broad gauge lines between there and Islamabad. Travel was a challenge, as obtaining a visa for Pakistan is far from straightforward and many of the participants did not receive their visa until just a few days before departure. In my case the authorities demanded proof of my pension, a bank statement, the ID card of the Pakistani tour operator and a utility bill to prove my address. All this just for a 30-day tourist visa! There was then the need for a PCR test for COVID-19 which had to be taken within 48 hours of departure and a booking for a lateral flow test 48 hours after my return home. Within 48 hours of arriving in Pakistan this latter stipulation changed to a booking for a PCR test 48 hours after returning to the UK. Worse still for the Americans in the group was the sudden demand by President Joe Biden that all arrivals in America had a PCR test just 24 hours before they landed home. The logistical challenges were surmounted through great teamwork, with those without a smart phone or computer helped by others in the group and we all navigated the requirements and made it home. A second aspect of the trip was the heightened security in Pakistan, as a result of the brutal murder and burning of a Sri Lankan man by Islamist extremists on 3 December 2021. Our group was escorted everywhere by 30–40 heavily armed anti-terrorist commandos and plans to stay in Attock City were blocked by the Interior Ministry because of security fears. This sounds quite frightening, but within 24 hours the group had developed good rapport with our protectors and I suspect the soldiers took more selfies and smart phone pictures during the week than we did. At one stage, our tour leader, Bernd Seiler, borrowed a soldiers Kalashnikov rifle and posed with the group, joking that he needed the AK47 to keep us all in order! Was it worth it? Definitely yes. The opportunity to visit a new country and photograph vintage British steam locomotives was unique, but the logistical challenges take their toll on all concerned and international travel looks like it will be a significant challenge for the foreseeable future.

Below: At 16.40 on 3 December 2021, HGS 2-8-0 2306 crosses the Attock Bridge with our charter freight as two local fishermen watch on. HGS stands for Heavy Goods Superheated and the 5 ft 6 in gauge 2-8-0s were built between 1912 and 1924 by a range of British and American builders. Our example was built by Kitson & Co Ltd. No.2306 was struggling to raise the boiler pressure and as a result, with the light fading, our train retained a diesel at the rear for this runpast, to provide a gentle push across Attock Bridge. The bridge is the most strategically important crossing of the mighty Indus River, between the provinces of Punjab and Khyber and is heavily fortified. The fort at the west end of the bridge can be seen on the left of this view and is mirrored by a similar building at the east end.
Photographic Details: Olympus EM-1 Mk III, 200 ISO, Olympus 12–40mm f2.8 13mm lens (26mm on 35mm), 1/200 at f8 (underexposed 1/3 stop).

HGS 2306 storms over Sakkardara Viaduct on the line from Attock City to Basal Junction. This spectacular route was opened between 1896 and 1897 and is an engineering marvel, with several large viaducts and multiple tunnels, which take the line through the very mountainous Kala Chitta National Park. **Photographic Details: Olympus EM1 Mk III, 64 ISO, Olympus 12–40mm f2.8 lens at 16mm (32mm on 35mm), 1/320 at f4.**

Above: On 4 December we had an alfresco lunch on the platform at Attock Khurd. In spite of the strict prohibition of alcohol for Pakistani citizens, we were able to source Murree beer, which has been brewed in Pakistan since 1860, when the company was set up by the British during the days of the Empire. Just before lunch, we had boarded our two guard's vans, to cross the Attock Bridge and see locations on the line towards Pesharwar. We were accompanied by a dozen commandos who happily posed for a picture as we rattled along through the Khyber countryside. It was quite a squash for 35 of us and our guards, but the trip over the bridge and through various tunnels was great fun! These three photos were taken with an iphone 12 pro max and the exposure details were not recorded.

Above: This image, taken on 8 December 2021, is a timeless scene and represents the type of photograph that inspired me to try and take more adventurous railway images. The progressive railway photographers of the 1960s, like Colin Gifford and Malcolm Dunnett, were part of a new wave of lens men who tried to move away from the standard three quarter front image of a train, where the quality was paramount, but the composition tended to be uninspiring. This approach, which included trying to set the train in the context of its surroundings, was continued into the 1970s by photographers such as Peter Cavalier, Geoff Silcock and Joe Rajczonek, who mainly photographed industrial steam. The Phoenix Railway Photographic Circle, founded by Wyn Hobson in 1971 to encourage this approach after steam had finished in Britain and of which I have been a member for 40 years, has also been an influence on my photography. This picture of HGS 2306 departing from Attock City towards Pesharwar has been rendered in black and white to better mimic the images that first inspired me to take up railway photography. **Photographic Details: Olympus EM1 Mk III, 200 ISO, Olympus 12–40mm f2.8 lens at 32mm (64mm on 35mm), 1/400 at f8 (on third stop underexposed).**

Right: From left to right are the author, Ralph Montagu, Dave Hill and John Birkinshaw, taking tea at dawn by the lineside at Attock Khurd. The adjacent Attock Bridge, which spans the Indus River and is illustrated separately, was opened in 1893 and remains one of the most important strategic crossings of the Indus between the Pujab and Khyber provinces. **Photographic Details: Taken with an iphone 12 pro max.**

2022

Having made it to Pakistan at the end of 2021, allbeit with PCR tests, vaccine certificates and multiple form filling, travel gradually became easier during the first half of 2022. Finally in June 2022, after three visits to the USA, COVID testing to enter the USA was abolished. As I write this at the end of June 2022, it looks as if international travel may be getting back to normal, although there is some chaos to be endured at airports. As far as photography is concerned, most of my efforts during the first half of this year have been to take images for the third edition of my "North American Railyards" book and hence the three visits to the USA. The year started with a now customary few days on the Settle and Carlisle line staying with Martyn Fordham. A couple of trips to Austria reassured us that international travel was relatively easy and so in March I headed to Southern California with Steve Arthur with the aims of visiting a few marshalling yards and photographing the wonderful Walong Loop on the Tehachapi incline using a drone. My third trip to the USA in June was arranged to coincide with what could be the last main line run for a Chinese QJ 2-10-2 steam locomotive. QJ6988, owned by Henry Posner III, was given special dispensation by the Federal Railroad Authority to run from its shed in Newton, Iowa to the workshops in Silvis, Illinois where it may be fully overhauled if funds are available. Its boiler certificate had expired a couple of years ago, but with the help of an expert crew of volunteers, the locomotive was prepared over the preceding week to run the 150 miles to Silvis. It hauled revenue earning freight as well as two passenger cars, totalling 2200 tons and the crew all had enormous fun running it flat out - the sound was amazing. As an added bonus, a Sunday outing from Silvis to Bureau Junction and back with a delicious barbeque on the lawns alongside the turning "Y" at Bureau Junction was organised. This was ostensibly to turn the locomotive! The images for 2022 have been selected to reflect what I think may be the future of railway photography. The two pictures taken on the ground were captured with an iphone, whilst the Walong Loop was taken with a drone.

Below: This image was taken at Hellifield station and is perhaps a harbinger of the future. Captured with an iPhone 12 Pro Max in poor late evening light, it highlights not only the quality of smart phone sensors, but also the powers of computational photography. The phone automatically selected "high dynamic range", sensing the flat light and photoshop has been used to further enhance the sky, bringing to life what would essentially have been a black and white image. On 25 January 2022, 70815 accelerates south with the Carlisle to Chirk log train, whilst 70813 waits for a path north with a Mountsorrel to Carlisle ballast train. The drivers of the two trains had swapped over at Hellifield, because this avoided them both having to stay in hotels overnight before they returned home. In a nod to the traditions of the past, the driver of 70815 has just been given a complementary cup of tea by the staff of the excellent tea shop (24H) on the station at Hellifield. **Photographic Details: iPhone 12 Pro Max, ISO 50, 5.1mm (26mm on 35mm), 1/120 at f1.6.**

Above: QJ6988 is readied for the 150-mile dash to Silvis outside the shed at Newton, Iowa on 11 June 2022. **Photographic Details: iphone 13 Pro Max, ISO 50, 5.7mm (26mm on 35mm), 1/4 at f1.5.**

Below: A trio of UP high horsepower diesels head north up the Tehachapi Pass with a double stack train which is captured traversing the fabled Walong Loop on the morning of 30 March 2022. Of course the sun went in at the last minute - some things never change for railway photographers! **Photographic Details: Mavic 3 Pro, ISO 100, 12.3mm (24mm on 35mm), 1/2000 at f2.8 (1 stop under exposed).**

Also by Michael Rhodes

THE ENCYCLOPAEDIA OF 21ST CENTURY SIGNAL BOXES

The definitive illustrated record of all Network Rail mechanical and heritage signal boxes that survived on the British railway network into the 21st century.

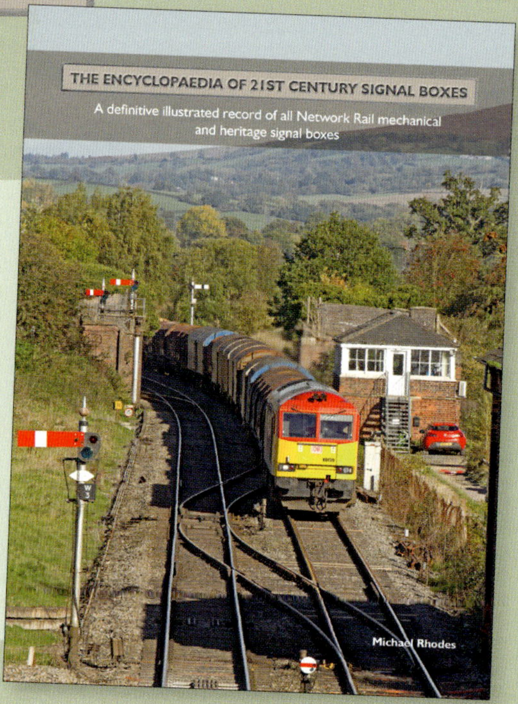

Contains historical data and photographs of every mechanical and heritage signal box. The first half of the book is a detailed survey of the years 2000 to 2018, examining all the signal boxes that closed in each year. Includes the date of construction, commissioning railway company and brief histories and anecdotes, plus interior and exterior illustrations.

The second half looks at all remaining signal boxes that were still in use in 2019, many of which will continue to see service for many years to come. Chapters are arranged by geographical area and contain details of the location of each box as well as descriptions of the nearest public place from where each box can best be seen. Every box is illustrated, many with both exterior and interior views.

An important and invaluable reference work. 320 pages. Hardback.

Cover Price £45.00. Platform 5 Mail Order Price £40.00.

FROM GRIDIRON TO GRASSLAND: THE RISE AND FALL OF BRITAIN'S RAILWAY MARSHALLING YARDS

Marshalling yards have been a pivotal part of the railways for nearly two centuries. The 1955 Modernisation Plan envisaged a huge expansion in wagonload freight and set in motion a massive programme of investment in new and refurbished yards. In reality, traffic began to decline steadily and few yards saw traffic anywhere near predicted levels.

Michael Rhodes traces the development of the marshalling yard and examines technical advances made in wagon control systems. He examines British Railways' policy towards marshalling yards in light of the 1955 Modernisation Plan and looks at every yard in detail; from the earliest yards to the first mechanised hump yards, the Modernisation Plan yards and finally Britain's last, fully automated, hump yards. A final chapter gives a global perspective and reveals the marshalling yard is still alive and well in other countries around the globe.

This definitive account is the result of over three decades of research. Contains over 500 photographs, 50 plans and 80,000 words. 288 pages. Hardback.

Cover Price £39.95. Platform 5 Mail Order Price £34.95.

ORDER BY POST, TELEPHONE OR AT www.platform5.com

Mail order prices are available from www.platform5.com or the Platform 5 Mail Order Department.

POSTAGE & PACKING: 10% UK; 20% Europe; 30% Rest of World. Cheques payable to Platform 5 Publishing Ltd.

Mail Order Department (50YP), Platform 5 Publishing Ltd,
52 Broadfield Road, SHEFFIELD, S8 0XJ, ENGLAND. Tel: 0114 255 8000